Early in the 1900s: the Wabash Bridge, photographed by the architect Ernest Boyer.

Dedicated to Richard M. Scaife

who provided the vision, risk capital, and faith needed to initiate our inner-city neighborhood restoration programs and to create Station Square, an urban center under the bridges along the Monongahela River.

Station Square, from the Smithfield Street Bridge.

Pittsburgh's Bridges

Architecture and Engineering

Walter C. Kidney

Pittsburgh History & Landmarks Foundation

Published by the
Pittsburgh History & Landmarks Foundation
One Station Square, Suite 450
Pittsburgh, PA 15219-1134
www.phlf.org
(412) 471-5808 fax (412) 471-1633

Walter C. Kidney, Author
Louise King Sturgess, Project Director and Editor
Greg Pytlik of Pytlik Design Associates, Designer

Manufactured in the United States of America

Library of Congress Catalog Card Number: 99-75974
ISBN 0-916670-21-X

The printing of this publication was made possible in part by the Revolving Fund for Education of the Pittsburgh History & Landmarks Foundation, established in 1984 through a generous grant from the Claude Worthington Benedum Foundation and later augmented through major grants from The Grable Foundation, The Mary Hillman Jennings Foundation, the Richard King Mellon Foundation, and an anonymous donor. Organizations and individuals who contributed specifically to this publication are acknowledged on page v.

The Pittsburgh History & Landmarks Foundation was established in 1964 as a non-profit historic preservation organization serving Allegheny County. It is dedicated to identifying and preserving the architectural landmarks, historic neighborhoods, and historic designed landscapes of the Pittsburgh region and to educating people about this region's architectural heritage and urban and landscape design history.

Please contact Landmarks for further information on its education programs, preservation services, and membership opportunities.

Pittsburgh's Bridges: Architecture and Engineering was typeset in Utopia, Bodoni, and Franklin Gothic Condensed and printed on Sappi Patina Matte 80 lb. text by Herrmann Printing & Litho, Inc. Beth Buckholtz and Diane Goodwin assisted in the design of the book.

The bridges as bearers of light: celebrating the lightbulb's fiftieth anniversary in 1929.

Contributors

The Pittsburgh History & Landmarks Foundation thanks the following for contributing to the publication of *Pittsburgh's Bridges: Architecture and Engineering.*

Benefactors

The Mary Hillman Jennings Foundation

PNC Bank Charitable Trust Committee

Patrons

Michael Baker Corporation

Anne L. and George H. Clapp Charitable and Educational Trust

Duquesne Light Company

HDR Engineering, Inc.

Jendoco Construction Corporation

MotivePower Industries

Marjorie M. Richardson and Sherwood Richardson, in memory of George S. Richardson

Sargent Electric Company, on behalf of "Light the Bridges"

Patricia Thauer Fund of the Pittsburgh History & Landmarks Foundation

The Walden Trust

Donors

Wilda Wible Aiken

Charles Covert Arensberg

Conrad C. M. Arensberg

Andrew Bell Armstrong

Roger Beal, in memory of Denton Beal

Charles H. Booth, Jr.

Cathy Broucek

Carl Wood Brown

Mr. & Mrs. John P. Davis, Jr.

Susan M. Devine, for Edward J. Halo and Richard W. Devine

Mary DeWinter Fund of the Pittsburgh History & Landmarks Foundation

Lu Donnelly

Mr. & Mrs. D. P. Dozzi

Thayer H. Drew

Gannett Fleming, Inc.

Dr. & Mrs. William S. Garrett

Geoffrey H. Goldberg

Harry C. Goldby

Philip B. Hallen

Torrence M. Hunt Fund for Special Projects of the Pittsburgh History & Landmarks Foundation

Monna J. Kidney

Jeffrey S. Lemley

Middy and Jack Markowitz

Pamela K. McCarter

Suzy McKain

Jack A. McKinnon, Jr.

Ruth and David Miller

Thomas D. Mullins

Mary Beth and Thomas V. Pastorius

Pittsburghscape/Joel Levinson

Marirose and John Radelet

Sally and Loren Siegel

Frank L. Stanley

Louise and Martin Sturgess

David J. Vater

Mary Wohleber

Alice and Hugh Young

Emma Ziegler Fund of the Pittsburgh History & Landmarks Foundation

Acknowledgements

One of the pleasures of writing a book like this comes from other people's willingness to help. Bridges form an attractive subject, and even the drab photographic records of a construction project have a certain thrill about them that conscious photographic artistry may not. Archives may show beautiful things coming to be, and their keepers are aware of this.

For graphics and information, we thank the following organizations and individuals:

County of Allegheny: Thomas G. Kelley, senior administrator of the Department of Property and Supplies; Denise Duncan and Margaret Stanley, Office of Graphics and Photography; and Charles Azzaro, senior records analyst, Office of Public Works, Department of Engineering and Construction.

American Bridge Company: Michael Cegelis, senior vice president, who gave us the opportunity to review the company's photographic files.

Archives of Industrial Society, Hillman Library, University of Pittsburgh: Miriam Meislik, associate archivist, in charge of the Pittsburgh City Photographer Collection.

Carnegie Library of Pittsburgh, Pennsylvania Department: Marilyn Holt and her staff.

Carnegie Library of Pittsburgh, Science and Technology Department: Malcolm Lindsay, a volunteer who is involved in a project describing the bridges of Southwestern Pennsylvania.

David Coyne, for his valuable and abundant volunteer work in photographing many of the bridges.

Christine Davis of Christine Davis Consulting, Inc., for information on the Monongahela Connecting Railroad Bridges.

HDR Engineering, Inc., in Pittsburgh, for giving us a copy of George S. Richardson's memoir, "History of Allegheny County Bridges," and for Beth Casey's assistance in letting us review the photographic files.

Pennsylvania Department of Transportation, Bridgeville: Patricia Remy, environmental manager.

City of Pittsburgh, Department of Engineering and Construction: Fred Fischer, project manager for bridges and structures, and Mark Stem, project engineer.

Pittsburgh Post-Gazette: Lizabeth Gray, editorial department.

U.S. Army Engineer District, Pittsburgh, Corps of Engineers: Conrad Weiser, Planning Division, and James A. Purdy, Natural and Cultural Resources Branch.

For reading the book in manuscript, we thank:

Laurie Cohen, M.L.S., Collection Services Librarian, Hillman Library, University of Pittsburgh

Steven J. Fenves, Professor Emeritus, Department of Civil and Environmental Engineering, Carnegie Mellon University

Sherwood Richardson, civil engineer, retired.

Illustration Sources

The numbers given below refer to pages on which illustrations from each source are found. All illustrations not listed below are from the photographic and library collections of the Pittsburgh History & Landmarks Foundation.

Allegheny County Division of Photography: 4 (bottom), 14 (bottom), 50, 57 (top), 94, 117 (top), 126, 162, 166, 167 (bottom), 169, 222 (also shown on back cover)

American Bridge Company: 142

Art Work of Pittsburg, The W.H. Parish Publishing Company, 1893: 13, 18, 46, 109 (top), 116, 179, 199

Art Work of Pittsburg, George E. White Company, 1899: 8, 37, 195

The Bridges of Pittsburgh, Joseph White and M.W. von Bernewitz, Cramer Printing & Publishing Company, Crafton, Pa., 1928: 31 (top), 51, 98, 103 (top), 115

Carnegie Library of Pittsburgh, Pittsburgh Photographic Library: iv, 7, 10, 15, 20, 25, 35, 40, 45, 54 (left), 56 (right), 95, 102, 104, 106, 109 (bottom), 111, 113, 143, 145, 146, 147 (top), 148 (left), 174, 186 (left), 187, 200 (bottom)

City of Pittsburgh, Department of Engineering and Construction: 198 (top right)

City of Pittsburgh, Department of Public Works, Bureau of Engineering: 204

David Coyne: 19, 27, 28, 29, 38, 48, 52 (left), 101 (bottom), 107 (bottom), 110, 114 (top), 121 (right), 124 (left), 130 (bottom), 131, 134, 135, 136, 137, 155 (top right), 156 (top), 164, 172 (top), 180, 181, 198 (top left), 205 (top left and right and bottom right), 208, 216 (top left and right and bottom left), 217 (top), 218, 219

Dictionnaire de L'Architecture Française, Eugéne-Emmanuel Viollet-le-Duc: 178 (left)

HDR Engineering, Inc.: 11, 163, 165, 224, 228

Clyde Hare: viii, xiv–xv, xvi, 5, 12, 16, 17, 59, 60, 62 through 90, 120, 175, 176, 177

George D. Hetrick Photographic Collection of the Pittsburgh History & Landmarks Foundation: 100 (bottom left)

History of Bridge Engineering, Henry Gratton Tyrrell, Chicago, 1911: 155 (bottom)

Jim Judkis: front and back covers, i, ii, x, 3, 6, 47, 108, 123, 140, 157, 158, 159, 188, 189, 197, 198 (bottom), 225 (top left), 235

Notas Interesantes Acerca de Pittsburg, 1889: 97

PennDOT: 170 (top and bottom left)

Pittsburgh City Photographer Collection, Archives of Industrial Society, University of Pittsburgh: xiii, 9, 22, 23, 24 (top), 44, 119, 127, 172 (bottom), 173, 182, 183, 184, 201, 202, 203, 206, 207, 213 (left), 216 (bottom right), 226

Pittsburgh Post-Gazette Archives: 48–49, 52 (right), 101 (top), 117 (bottom), 121 (left), 128, 129, 132, 133, 138, 139, 148–49, 160–61, 170 (right), 171 (left), 193, 205 (bottom left), 220, 221

William Rydberg, PHOTON: 33; from the collection of the Allegheny County Division of Photography: 122, 124 (right), 141 (bottom), 153, 154, 155 (top left)

Supplement to *Engineering News,* June 22, 1899: 194 (bottom), 234

U. S. Army Engineer District, Pittsburgh, Corps of Engineers: 58 *(Allegheny River Navigation Charts),* 103 (bottom), 118

The Eliza Furnaces of Jones & Laughlin and the Brady Street Bridge over the Monongahela River. February 1975.

Contents

St. Josaphat's on the South Side Slopes and the bridges beyond.

Introductory Note

Without bridges, the Pittsburgh region would be a series of fragmented valleys, hillsides, river plains, and isolated communities. With them, we come together to form a mighty unity.

Some 1,600 bridges bind our city and county together, making places accessible that otherwise would not be accessible and resulting, in many cases, in ingenious engineering structures full of color and drama. In *Pittsburgh's Bridges,* Walter Kidney reminds us that our region has been and *is* a place of extraordinary architecture and engineering.

In dedicating this book to Richard M. Scaife, we recognize his deep commitment to the Pittsburgh region. Through his philanthropic efforts, he has improved the lives of many people and strengthened organizations dedicated to the arts, education, health, and human services.

Decades ago, he envisioned how our region could be strengthened through the preservation and beneficial use of our landmark architecture, engineering monuments, and historic landscapes. Through the Sarah Scaife and Allegheny Foundations, Richard Scaife granted the first funds for the Pittsburgh History & Landmarks Foundation's inner-city neighborhood restoration efforts during the 1960s and '70s, when our historic preservation programs were unique, untried, and risky. Since 1964, Dick has gone with us on a regular basis into the neighborhoods to talk with residents, observe their work, and discuss community needs.

When we objected in the 1960s to the destructive urban "renewal" policies that demolished the center of Pittsburgh's North Side and threatened the Allegheny Post Office, Dick gave us the lead grant to save the 1897 landmark. Now more than 100,000 people visit the Old Post Office building each year, in its new life as the Pittsburgh Children's Museum.

Beginning in 1975, thanks to Dick's leadership and vision, the Allegheny Foundation gave Landmarks the risk capital to begin Station Square, a riverfront project combining the adaptive use of five historic railroad buildings with new construction. The 52-acre site on the south shore of the Monongahela River eventually extended over a mile, from the Panhandle Railroad Bridge, underneath the Smithfield Street Bridge, past the piers of the former Wabash Bridge, under the dramatic Fort Pitt Bridge, and past the remaining abutments of the former Point Bridge.

As the prime developer of Station Square from 1976 to 1994, Landmarks was able to put its urban planning principles into action. When Landmarks sold Station Square there were 3,000 jobs, 134 businesses, and a series of attractions that brought 3,000,000 visitors a year to the site. Each day hundreds of people cross the Smithfield Street Bridge, linking downtown to Station Square. Thanks to Landmarks' initiative, the upper chords of the trusses are lit with 313 lights so the bridge is as distinctive at night as it is by day.

With the sale of Station Square, Landmarks created an endowment for historic preservation in the community, thus fulfilling one of the goals established by Dick at the outset of the Station Square project.

When we think of our abundant bridges, we think of great designers and their public and private patrons; we reflect on all those who are linked together to work for a finer city. And we think of Richard Scaife, whose faith in the power of historic preservation has enabled Landmarks to create new life and opportunity for our region.

Historic preservation, like our bridges, weaves through our community in abundant and often unexpected ways. It is itself a bridge between the past and present, carrying us into the future.

Arthur P. Ziegler, Jr.
President, Pittsburgh History & Landmarks Foundation

Preface

The Bridge is a cliché. Metaphors employ the bridge so often that it is as though the wordplay preceded the object. People strive to "bridge the gap" and "span the divide." Bridges take us from one musical movement to the next. They replace the function of lost teeth. "Bridge to the future" is common in political rhetoric.

Concretely, bridges are a luxury. Unlike Food, Clothing, and Shelter, the bridge is not a requirement for existence. In the modern world, however, bridges seem to be a necessity. They are a vital part of our infrastructure and our way of life. Without our bridges, the economy would halt. Goods and services would not be distributed. War clearly demonstrates this phenomenon. Bridges are targeted and destroyed to prevent the enemy from advancing, but afterward their absence hinders relief aid for the stricken. Luckily, most of us can take the bridges we use for granted.

Unlike any other form of architecture, the bridge symbolizes just how far our civilization has developed. Ever-longer spans are borne of humanity's ever-expanding mastery of the natural world.

At a conceptual level, the bridge represents a step towards completion. Problems are incomplete, and solutions displace the incomplete with the complete. The bridge represents, at its core, the desire of civilization to complete itself.

Arthur Lubetz

Arthur Lubetz Associates, PC

The Pittsburgh & Lake Erie's Beaver-Monaca Bridge is opened: May 14, 1910.

The surprise of Pittsburgh, upon exiting the Fort Pitt Tunnel.

On Bridges

An engineer's career is spent on the bringing about of non-events. Take our present subject, bridges. If we are driving, a matter of seconds will usually take us from end to end of a bridge, and we need not heed what is to either side or below. We need not wait for a ferry, scurry down and up valley slopes, exert ourselves in any extraordinary manner going along the roadway of a bridge, which is simply a stretch of roadway like any other.

Positioning that stretch of roadway, giving it safe support, is of course a process that can occasion severe and manifold difficulties. The void the bridge crosses, the static or moving elements that occupy the void, the matter that constitutes its walls: understanding what is to be crossed is fundamental. So is an understanding of structural materials and their systems of assemblage so as to create a dependable bridge structure. The approaches of a bridge, as planned, accommodate or frustrate traffic, improve or degrade adjoining neighborhoods. Finally, the engineer and the lay people who influence the design of a bridge may or may not create, in the bridge, a poem of curves, a heroic essay in straight lines, a no-brain affair of beams set down on bents, or trusswork that seems merely the constructional equivalent of fine print, without lucidity.

Main spans and approaches along the Monongahela River.

A bridge for the automobile, called for in 1910, opened in 1928, the year of this picture. Beyond the Liberty Bridge, the motorist could go through Mount Washington via the Liberty Tubes to the South Hills, or up the hill via the Mount Washington (now P. J. McArdle) Roadway.

To Homestead on the High Level Bridge.

A bridge that ignored conditions on the void it crossed: the towboat *Duquesne* confronts the Sixth Street Bridge in 1914. The Allegheny River was flooding, admittedly (left).

A poem in the making: Jack's Run Bridge on Ohio River Boulevard. September 8, 1930 (below).

A utilitarian structure: Jones & Laughlin's Hot Metal Bridge, with the Eliza blast furnaces hard at work in the mid 1960s.

In the Pittsburgh area, settlers were faced with the complicated consequences of the terrestrial crust's floating around, a great westward-flowing river that formed a delta where Pittsburgh now is, events in plate tectonics that cut off the river, the formation of new rivers, and two Ice Ages that shoved the rivers around through glacial action. Settlers established a trading post, a military establishment, a county seat and market town, a river port, a growing chain of industrial towns, and a conurbation of neighborhoods, all draped over plains, hills, and ravines with differences of level as great as 500 feet, and divided by rivers. Development obviously meant bridges, 50 feet above the water, 100 or 200 feet above the valley floors, dominating the landscape eventually as other contrivances for the purpose, tunnels and inclines, could not.

The graceful Washington Crossing Bridge in 1996, seen beyond the roofs of Millvale (above).

The Bloomfield Bridge, 1914–80, a rawboned cantilever construction (opposite).

CARMAN-ROBERTS C

The High Bridge, between Pittsburgh and Bellevue, was up by 1893. Built for trolleys, it had to accommodate fast-moving and rather heavy loads, yet its system of short spans and towers and its deck level 100 feet above the valley floor made it look light and airy (above).

The houses of Junction Hollow and the serene arched span of the Schenley Bridge: June 23, 1914 (opposite).

B. of E.- JUNCTION HOLLOW BRIDGE, SCHENLEY PARK. 6-23-14*4141

A bridge is a means of connecting separate places on the same level that can be crossed absent-mindedly, unnoticing. It also can be a handsome work of structural art or a piece of visual rhetoric, granted. But there are other ways of perceiving it. Sonically, a bridge can be a living thing. You are standing on a river shore, and far above you is a bridge, with traffic briskly passing over an expansion joint: pa-*pum*, pa-*pum* . . . pa-*pum* . . . pa-*pum* pa-*pum* pa-*pum*,

and so on. You are in an excursion boat, approaching then passing under a bridge, and the echoes of the parted waters and engine beat go into a crescendo, then a diminuendo, from the bridge's floor and piers. A bridge, too, is a means of screening a distant landscape, cutting off a long perspective, controlling through its structural work how much you see and dividing it into fragments of color and geometry. Crossing a valley, it imposes form and scale, a shape to the void that states where the walled space ends and the sky begins. Finally, a great bridge, high above a valley, can create a wonderful feeling of There versus Here. You are on a valley floor, looking up at the stately row of illuminated bridge lamps, showing the way between places not seen. Or you are standing on a bridge on a summer evening, looking at a little neighborhood 100 feet below, hearing the shouts of children whose running forms appear under a corner street lamp, a world away down there.

A frame to a space: the Panther Hollow Bridge (above).

A hinge and a house: contrasting scales in Homestead, under an approach to the High Level Bridge (opposite).

The Monongahela River in the 1980s: five distinct structural types in about one-and-a-half river miles (opposite).

The Monongahela in 1893: plain, functional spans, with a little attempt at artistry at the portals (above).

The Allegheny River in 1874, with John Augustus Roebling's Sixth Street Bridge to the left of center, and the wooden Ninth Street Bridge to its left.

Twenty-four reasons for building a bridge: maquette by Frank Vittor for a memorial to victims of a ferry sinking near McKees Rocks (left).

Bridges against the landscape: *Industry's Increase* by John Kane, with the South Tenth Street Bridge.

Framing of sky and land by the West End and Birmingham Bridges (left and opposite).

Marlboro Lights

KAUFMANNS
CLOTHING
BETWEEN
E. JORDAN

The Bridge as Architecture

In the Pittsburgh area from the start, there were attempts to introduce some element of visual art to the structure of a bridge: not in every case, to be sure, and in a good many other cases it appears that the builder or engineer had no real heart for the attempt. Until 1896, river bridges around here were private and charged tolls, and the punctuation of the crossing with portals had some rationale, it may be, in implying stops. But there was long a tension, we may infer, between Engineering in its objectivity and Architecture, which sought to beautify in one way or another such a large and conspicuous object as a bridge. Functionalists of the 1920s might find *no* reason, ever, for the architect to modify the appearance of a superstructure, abutments, or any subordinate features. Some people might instinctively envision a parabolic arch taking a valley in one balletic leap, while forgetting the complicated but efficient trusswork of, say, the Panhandle or Hot Metal Bridge crossing our Monongahela River.

From a portal of the Manchester Bridge.

And it is truc that architects and other designers have often created additions to, rather than integrations with, the bridge structure. Returning for a moment to portals: a bridge needs some sort of cross-bracing to keep its structure from transverse distortion; but this, by itself, is

Gustav Lindenthal attempted to beautify the towers of his Seventh Street Bridge of 1884, on the Allegheny (opposite).

inconspicuous. It was an "architectural" compulsion that led to decoration of the entrances to "through" trusses and arches, typical of rivers, with: the weird quasi-Baroque carpentry of the Union Bridge (1874); the crude original quasi-medieval portals of the present Smithfield Street Bridge (1883); the Beaux-Arts piers of the Sixteenth Street Bridge (1923); or the giant staples that terminated the Point Bridge of 1927. None of these features contributed to their structures in any important way, if at all.

The Union Bridge, 1874–1907, with its quasi-Baroque portals.

Yet they did, of course, contribute to the public's impressions of the bridges. So did some other features that the bridges, as constructions, did not really need. The Liberty Bridge (1928) had mighty-looking piers, rising even above deck level but doing nothing above the spring lines of the cantilevers. The often-found presence of capitals and rustication cast in concrete bridges is another case in which "architecture" was imposed on engineering, or where engineers were trying to be architects. It almost seems as if we are left with a few details commonly expected to be decorative—piers and abutments to some extent, railings, lamps, plaques—that belong with the engineer's pure art.

Pittsburgh's Art Commission, formed in 1911, was not the initiator of the appearance-conscious City bridge, as the view dated November 11, 1910 demonstrates. The reinforced-concrete Meadow Street Bridge is like a grand and simple gesture, straight and level on its concrete arch, that crosses an eroded valley to the clutter of houses in Larimer. As was common in these early days of concrete in a public setting, unnecessary masonry forms, rustications, brackets, capitals, and arches over narrow voids, are applied to civilize the dead gray material. An attempt at beautification is being made.

The Bloomfield Bridge was strictly business as a whole, but its stark utilitarianism was mitigated by lampposts of Art Nouveau character (opposite). The designer was probably Stanley L. Roush.

Yet can we deny that trusswork is sometimes ugly and chaotic, that it obscures a view that an arch or catenary might reveal and beautifully frame? The expediencies of crossing a river might favor a series of trusses, with some sort of portal decoration as a sop to the aesthetes, but broader considerations might favor the arch or catenary as more articulate and graceful, and the client might be willing to sacrifice money to get one or the other.

Pittsburgh's Art Commission, coming to be in 1911, and Stanley Lawson Roush, designing for the City of Pittsburgh and Allegheny County into the 1930s, helped give this area handsome and lucid bridges at a time of very extensive construction that was characterized too by progressive engineering design. We are still living with the bridges of the 1910s, '20s, and '30s, though railings, lamps, whole bridges have sometimes departed. It is notable—and

The Hoeveler Street Bridge in the East End was treated by Roush as a masonry work of 14th-century Italian character: a watercolor rendering by Roush (above), and a 1913 construction shot (below).

meaningful?—that other dramatic local improvements occurred in this period: the Pittsburgh Civic Center in Oakland between 1895 and 1940 and a handsome series of Pittsburgh public schools between, say, 1910 and 1940. Through such forms of construction a picture at least of a new and better Pittsburgh became visible.

The Forty-third Street Bridge of 1870 lasted until 1924 (opposite).

Quotables

Here are some documents from our bridge history. The first is a memento from the period of the wooden covered bridge, which did not end locally until 1924, when the Forty-third Street Bridge came down. Many of the articles are from the great bridge-building periods of the 1920s and '30s. The last pieces quoted are from recent times.

Old Sweet Song

An attempt to utilize in a singular way [vibrations in wooden bridges] *is recorded in the* Proceedings *of Engineer's Society of Western Pennsylvania for 1892. When the old wooden bridge at Ninth Street was being removed in 1890, a local unnamed maker of fiddles asked for and was given a block of the timber from the old wooden arches. He made two violins from this wood upon the theory that the constant vibration in these timbers over a period of 50 years would be equivalent to many years playing a finished instrument. These violins were said to be unusually sweet toned and commanded a price rather high for those days. Who is now the possessor of these unique instruments?*

Joseph White and M. W. von Bernewitz
The Bridges of Pittsburgh

The Bridges of Pittsburgh

The City of Pittsburgh and the County of Allegheny are the largest owners of great bridges in the country, in proper proportion to their size, with the possible exception of the City of New York, which is built on an island and surrounded by large bodies of water. . . .

A bridge is said to represent the drama in building construction. There is something intensely dramatic and fanciful in the appeal of the bridge to all classes of people, under all conditions of nature. All traffic converges and concentrates on the bridges. They become a daily necessity and a familiar benefactor, giving convenient passage over some natural obstruction.

When the highway leaves its natural bed and soars through the air on thin and film-like construction, to convey the people, the trains, the street cars and all the traffic in safety across a wide river or deep valley and when we look over the railing of the bridge into the dark flowing water or the ground many hundreds of feet below, we are all deeply stirred in our emotions.

This accomplishment of man's handiwork is one of the most dramatic things in modern life. The bridges are a very vital part of the life of the community. It would be impossible to get along without them. Therefore, Pittsburgh has the opportunity to do well with this problem, to take a step in advance of what has been done in the past.

The bridges of Pittsburgh that are built and are to be built, do and will make the same dramatic appeal to all the people. . . .

As bridge design progressed in America, there has naturally come with it appreciation and demand for those bridges which enter into the panorama of a city striving to look well, a search for handsome, clean cut lines and noble appearance. The opportunity for durable bridges of economical design with the accompanying qualities of noble appearance and as much beauty as can be made to accompany the foregoing qualities, has been engaged by those in charge of the design of five recent major Pittsburgh bridges. . . .

In Pittsburgh many men have lent their voice and energy in advocacy of the best bridge design. The late Mr. John W. Beatty, Director of Fine Arts, Carnegie Institute, for many years worked and succeeded in securing the best talent for the improvement of the appearance of Pittsburgh bridges. He worked also for the improvement of the appearance of all public property and Pittsburgh will owe him a debt for his effort that can scarcely be repaid. One of the chief results of Mr. Beatty's effort in collaboration with other Pittsburghers, was the formation of the Art Commission of the City of Pittsburgh. The function of the Art Commission is that it has the power to prevent the construction of any properties of the City of Pittsburgh which do not contain good taste and appearance. The Art

Commission has control only over properties of the city on land owned by the city. The county bridges are on land owned by the city, the Federal Government and the County. These bridges are not entirely within the control of the Art Commission.

Since the appointment of Mr. Norman F. Brown, Engineer, and Director of the Department of Public Works of the County of Allegheny, a relation has grown up for the improvement and study of bridge design of a quality which cannot be too highly commended. Mr. Brown has been entirely responsible for this relation. Surrounded as he is by restrictions and particular requirements of all kinds laid upon him by the Federal Government, the River Interests, the Public Utilities, the various Planning Organizations of the City and County, and definite limits of cost, which have taken little into consideration from necessity, of any provision for any aesthetic quality or design, he has faced his problem in the broadest sense. He has invited the criticism of the members of the Art Commission and welcomed them to his conferences and sought their advice when he was but adding to the load which he was carrying.

The Art Commission has operated through a committee of the architect members in endeavoring to join with Mr. Brown in the same spirit. The architect members consist of Edw. B. Lee, Chairman, Wm. Boyd, Vice President, and Frederick Bigger. Mr. Stanley L. Roush, Architect, attached to Mr. Brown's organization, has been a member of these conferences. In addition, Mr. Brown has a force of engineers of great ability and joined with them has been freely given the assistance of Mr. Covell, County Engineer. . . .

The first development of this program has been done in the construction under one contract of the Sixth, Seventh and Ninth Street bridges. We have here a solution that is unique and is without doubt without any parallel anywhere in the world. The problem presenting itself was the construction of three bridges, a city square apart, on which the conditions were absolutely the same, resulting, as there was no reason for doing otherwise, in a simple group. These bridges carry, at comparatively low grade, the projection of city streets, across a narrow navigable river in the heart of a great city. There was no effort made for decorative portals, sculptural accessories or forced scale, but merely the projection of a roadway into the air and across the river with the least interference with the panorama and flanking buildings of the city. . . .

An urn by Roush, on the Beechwood Boulevard Bridge.

If a successful solution is found there will be a project developed that is almost new in history—that is, a real working relation between able structural bridge designers, and architectural ideas applied in proper proportion to some great Pittsburgh Bridges.

Stanley L. Roush and Edward B. Lee
"The City of Bridges—Pittsburgh"
The Charette, March 1925

The Art Commission Gets a Say

Early in 1921, when the construction of new bridges over the Allegheny river began to be discussed, members of the [Pittsburgh Art] *Commission initiated a quiet campaign to convince the authorities concerned that aesthetic as well as utilitarian factors should be considered in planning the proposed river-crossings. The County Commissioners quickly accepted this opinion, and in 1922 invited the Art Commission to pass upon the artistic merit of certain designs. . . .*

At no period in the history of the world has a community been called upon to erect so many major bridges at one time as metropolitan Pittsburgh is building today; and if by its contribution the Art Commission has assisted in making them dignified and beautiful, its existence has been justified.

Anon., "Beauty and Bridges"
The Charette, March 1925

Castings from the Manchester Bridge.

The Architect: Taking a Place

The first successful organized attempt to keep bad artistic work from being bought or built in the City of Pittsburgh was the Art Commission sponsored by Mr. John W. Beatty and active and able people interested in better things, about 1907. [The Commission was established in 1911.] *Mr. Beatty worked in and out of season as best he understood the situation and gave of his time and energy without stint. . . .*

By the Art Commission's influence an architect was attached to the Department of Engineering, Bureau of Public Work. This man was Stanley L. Roush, who made a success of his job and his opportunity. When the design of the 16th and 40th Street bridges, and also a prospective bridge at Clairton, came up, the County Commissioners were persuaded through their relations with the Art Commission and its members, to make a change in the customary procedure and put the bridges in charge of an Architect. . . .

The first two bridges were put in complete charge of Architects to design and supervise, including all engineering work, to be furnished by the Architect. The Clairton bridge was different. In this case the Architects were to consult with the County Engineer who was in charge of the bridges. In the case of the 16th and 40th Street Bridges the Architect had to secure the approval of the County Engineer.

The awarding of the design, including the engineering work to an Architect was badly advised. The Pittsburgh engineers took exception and offence at the taking of a class of work long looked upon by their profession as developed by and peculiarly belonging to them. It seemed to some of us that the engineers had grounds for their contentions. The engineers, we heard, had organized themselves and had prepared a plan by which they proposed to contest the awarding of any more bridges to Architects. . . .

Architect Benno Janssen, sandwiched between engineers.

It always seemed to some of us that this typical factional Pittsburgh method of fighting among ourselves about some good idea until the idea was lost in the turmoil of the fight, could and should be avoided. . . .

M. K., "Engineers and Architects"
The Charette, April 1924

The Architect: Knowing His Place

In the long debate between Architect and Engineer about the bridges and how they are to be done, the real issue has for many become much clouded. To the lay, and perhaps also to the Engineering, mind the Architect is pushing his cause for the quite understandable "quid pro quo" involved and for no other reason. This is a great injustice, albeit one that the Architect has brought upon himself.

Now the simple truth of the matter is that the Architect has but one real interest, to wit, to who: that the final result may be bridges good after their kind, as handsome as may be under the limitations of site and function and if this result is attained the manner of its attainment is not only a point of minor moment, but in very truth of no moment whatever before the eternal gods.

The only justification the Architect has for commending himself to the attention of the political powers in whose hands lie the direction of this work, is that idealism that looks to the final result to show that crowning element of beauty which can come only from the hand of him whose native instinct, training, experience and ideals coincide with those of the Architect.

Excellent results have at times come from the hand of the Engineer working along purely practical and utilitarian lines, but speaking broadly good design comes only from the conscious and definite effort to refine and beautify the structural forms that the nature of the problem demands. The Architect makes no claim for the entire charge of this work; he would be as competent to that task as the engineer; but the nature of the problem calls for the friendly and intelligent collaboration of the two branches of the profession, in every respect one and the same except that the one branch has specialized on construction and the other on design.

"Bee," "Fiat Lux"
The Charette, April 1924

Credit where credit is due.

The Wise Politician

It was my understanding that in appointing Norman Brown as Director of Public Works, Brown had insisted that he would have an absolutely freehand in the selection of his staff. Supposedly [County Commissioner Joseph Armstrong] *told him "I would not have it any other way. I know this is the only way you can do your job. I want you to know, however, that after you get through naming your appointees I may add some others. These people will not disturb you. You may never see them."*

George S. Richardson
"History of Allegheny County Bridges" MS., 1979

Red Tape in 1928: Getting Plans for a River Bridge Approved

Briefly summarizing, after the approval of the United States Congress, the War Department, State Water and Power Resources Board, the Pittsburgh Flood Commission, the County Planning Commission, the City Planning Commission of Pittsburgh, the Citizens Committee on City Plan, the Pittsburgh City Council, the Art Commission of Pittsburgh, the Public Service Commission of Pennsylvania, and the Grand Jury and Court of Quarter Sessions, the plans are submitted to the County Commissioners for approval. They then refer the plans and specifications to the County Controller, whose engineer reviews the plans before the contract is finally advertised.

Joseph White and M. W. von Bernewitz
The Bridges of Pittsburgh

Mandated Drabness

The builders of our city understood the importance of constructing handsome bridges. From the first county bridge in 1819 [really 1818] *to bridge builder John Roebling's Sixth Street Bridge, an enormous amount of attention was given to the unique gracefulness and openness of our bridges. They were considered the Smoky City's saving grace at a time when it was thought to offer nothing architecturally attractive.*

Pittsburgh's bridges are lasting testaments to how Pittsburghers conquered this area's topographical limitations with architectural splendor. . . .

[The] *county engineers demonstrated a profound understanding for the role attractive bridges must have in a region with at least as many bridges as Venice, Italy.*

Today, however, the City of Bridges is fast becoming the City of Neutered Bridges because state and federal bureaucrats are requiring that concrete barriers and cyclone fences be used on most renovation projects.

These ugly slabs of concrete are fastened to the deck and traverse the bridge's expanse. These sterile blockades obscure the traveler's view from numerous area bridges and ruin the carefully planned style and spirit that earlier engineers toiled to achieve.

Bureaucrats . . . contend the barrier is supposed to be a safety feature. They also say the fence is designed to prevent people from dropping objects from the bridge. However, when these modern-day "bridge safety measures" are used, Pittsburgh ends up with another unsightly bridge.

State and federal officials, who control the purse strings to essential bridge-renovation dollars, are unlikely to approve funding for projects that do not use them. . . .

Frank Lucchino, Allegheny County Controller
Pittsburgh Post-Gazette, August 27, 1988

The original Liberty Bridge detailing.

The Liberty Bridge after reworking to new standards in 1983: railings gone, original lampposts gone, towers at the piers cut down.

An Exchange of Letters

Robert J. Gangewere to Alex J. Sciulli, Director, Department of Engineering and Construction, City of Pittsburgh, December 20, 1992:

The executive committee of Preservation Pittsburgh has asked me to write you about the renovation of the Schenley and Panther Hollow Bridges in Schenley Park, urging you towards a higher degree of concern about the final appearance of these bridges. . . . Two currently planned aspects of the renovation will effectively ruin the traditional appearance of the Schenley Park bridge for both pedestrians and motorists: the use of high, curving chain-link fences on the outside railing, and the installation of concrete Jersey Barriers along the bridge curbs.

Clearly the Jersey Barriers designed for 60-mile an hour turnpike traffic have no place on a 25-mile an hour road into a park, a road which was designed as the major public promenade between a museum and flower conservatory, and a road which today gives thousands of students access by foot to the two universities, the library and museums, and Phipps Conservatory. . . .

Chain-link fences are the second problem, since they create a no-view zone ahead of and behind users of the bridge. Any effort you could make to limit the size of these fences would be greatly appreciated—if in fact you cannot eliminate them. . . . State and federal regulations which have no regard for local aesthetic and social circumstances are not finally the best solution for every Pittsburgh bridge project, and the preservation community urges you to seek the highest possible preservation standard for these bridges. . . .

Sciulli to Gangewere, January 19, 1993:

I am in receipt of your letter regarding the rehabilitation of the Schenley Drive and Panther Hollow Bridges in Schenley Park. You can be assured that the City of Pittsburgh is very concerned about historic preservation, especially in and around Schenley Park. This has been demonstrated by the fact that the City was willing to spend an additional million dollars to rehabilitate rather than replace these two structures. Unfortunately, this must be balanced with the need to protect the traveling public and the City's liability in an increasingly litigious society.

The first of two items which you refer to in your letter is the concrete Jersey Barriers required to be constructed between the sidewalk and the roadway. While it is true that the posted speed limit in the park is 25 mph, I think you'll agree that traffic often travels at speeds which greatly exceed the posted limits. These barriers are specifically designed to deflect errant vehicles back onto the roadway surface, as well as protect the pedestrians on the sidewalk, regardless of the running speed of the vehicle. . . .

Chain link fences are identified as the other problem with the bridge design. Please note that this fencing is only required on the Schenley Drive Bridge and will not be installed on the Panther Hollow Bridge. The fencing is required by the railroad operating beneath the bridge and has been ordered to be installed by the Pennsylvania Public Utility Commission, which also has jurisdiction over this project. A private parking lot, as well as a City street beneath the bridge, will also be protected by this fence. On more than one occasion the City has received complaints about objects being thrown from the bridge. The fence is the minimum protection which can be provided. In order to reduce the amount of fencing we have limited its construction to the main span only. . . .

Since F.H.W.A. and PennDOT are providing the majority of funding for these projects, they, as well as the City, may be held liable for any accidents which might occur as a result of these standard safety features being eliminated. . . .

The City of Pittsburgh appreciates your concerns and is committed to preserving the historical features of this area. However, to continue to use these bridges we must, in order to protect the interests of all concerned parties—as well as the traveling public, incorporate current standard safety features.

The Schenley Bridge c. 1995: once a gracious feature of a park, now more like a cattle run.

Bridge Anatomy

The Pittsburgh area has been called a bridge museum. We have no need for the bascule, swing, or lift bridges so prominent in Cleveland or Newark, but we have most of the other types normally in use. In addition, we have a few definite oddities, such as the Smithfield Street Bridge or the tufa-faced arches in Schenley Park (see page 200).

The Thirty-third Street Bridge from the Strip to Herron Hill, late 19th century.

Leaving aside railroad bridges and the canal viaducts that once crossed the Allegheny, our bridges are intended to carry roadways. The roadways rest on a light structure called a *deck,* and the deck, for the most part, on a heavier structure of beams called a *floor*. The part of the bridge that actually spans the void created by the river or the valley is called a *superstructure*. This rests on constructions at the edges of the void called *abutments,* and sometimes on constructions in between. If such intermediate constructions are solid, concrete or masonry, they

Here is a Pratt truss from the late 19th century, location not known (opposite). In the Pratt truss, thin diagonal tension members descend across single panels to the feet of posts that are in compression. Here, the posts are built up of thin pieces riveted together in a way that prevents them from flexing under pressure. The portals and upper chords are in compression; the lower chord, made of eyebars, is in tension.

are called *piers*; if they are squarish portal-like frameworks running from side to side they are called *bents*; and if they are squarish in plan they are called *towers*.

Bridges are usually defined according to their superstructures. The part of a superstructure between any pair of supports is a *span*, and most of our short-span bridges function as beams. Such bridges are often very simple: several *girders* or *stringers* (depending on size), lengthwise beams usually of steel or reinforced concrete, are set on abutments and perhaps on bents or towers to carry the bridge floor. Ramps are often made this way, and so are overpass bridges. *Plate-girder* bridges often carry railroads, or roads that pass over railroads. A plate girder is an oversized girder, built up from easily rolled parts: a plate of steel or, earlier, wrought iron set on edge, a *web*, assembled with *flanges*, projecting edges that strengthen it.

A plate-girder bridge of 50-foot span over Montour Run. Late 19th century.

A footbridge in Emsworth, late 19th century.

A Pratt-truss bridge over the back channel of the Allegheny at Herr's Island.

The *truss* is usually a sophisticated version of the beam, used where spans are fairly long. In any beam, the lower edge, or *chord,* is in *tension,* subject to forces that tend to pull it apart. The upper chord, on the other hand, is in *compression,* subject to forces that tend to crush it. A truss restrains both forces, partly by the thickness of its parts and partly by causing the forces to follow certain paths. Any genuine truss is based on the fact that the angles of a triangle cannot be changed if all three sides remain connected and unaltered. Though there are many kinds of trusses, every truss is a composition of triangles, some of whose sides are tension members, some compression members, and some one or the other depending on loading conditions. The complexity of a large truss is due to the peculiar loads it must meet. If a train starts to enter a double-tracked railroad bridge, for example, it loads one corner of the first span much more heavily than the other three corners are loaded. The trusswork that keeps the span from being twisted out of shape lies within the main supporting trusses, within the floor and sometimes above the tracks, so that the bridge becomes like a hollow beam. Not only must off-center and moving traffic loads be met: even the wind is important. Gusts of wind can send shock waves through a too lightly built bridge that can tear it to pieces.

A scene of unknown location, with four through Pratt-truss highway spans in the foreground, and a railroad bridge beyond (opposite).

A span with an unusual elaboration of a Pratt truss for the Ohio Connecting Railroad across Brunot's Island, under construction in 1900. The supporting falsework is an engineering work in itself. The falsework maintains the form of the structure until the structure can support itself.

The Wabash Bridge (1904–48), with cantilevers complete and the suspended span under construction (left).

A bridge of 1889 across the Monongahela at Dravosburg (below): three Whipple trusses, one of 312 feet, two of 220 feet. As in a Pratt truss, a Whipple truss has diagonals in tension, but they cross two panels rather than one.

A truss that lies wholly below the deck level is called a *deck truss*. A truss whose lower chord supports the deck is called a *through truss*. A shallow through-truss bridge with unconnected upper chords is called a *pony truss*.

A *cantilever truss* bridge is even more sophisticated than the truss bridge that functions as a beam. A cantilever is an extended arm, whose upper part is normally in tension and whose lower part is normally in compression: a bracket on a large scale, in other words. The inner arms often support a beam-like *suspended span*, while the outer arms are firmly anchored. The Sewickley Bridge of 1911, the Homestead High Level Bridge, and the Liberty Bridge (despite its arch-like appearance) are all cantilever bridges. Together they show the variety possible in cantilever design.

Pittsburgh also has a number of examples of the steel *arch*, which can be found in the Schenley and Panther Hollow Bridges, the West End Bridge, and the Fort Pitt Bridge. The arch is quite simple in principle if not wholly so in practice. It receives forces that weigh on it, mostly in a vertical direction, and carries them on descending curves to places where they are finally held in check by some kind of support or tie. In most arch bridges, heavy, solid abutments absorb the *thrust* of the arch or arches. In a *tied-* or *bowstring-arch bridge*, the floor acts as a tie connecting the arch ends, so that the abutments resist only vertical loads. Naturally there are complications from off-center loads, and an arch used in a steel bridge will have trusswork somewhere to keep the arch from distortion. In the Panther Hollow Bridge the trusswork is in the *spandrels*, the roughly triangular members between the arch and the roadway. In the West End Bridge the trusswork is in the arch itself, which is in effect two arches, one inside the other. In the Fort Pitt Bridge, which is double-decked, the trusswork runs between the upper and lower roadways.

Here is a combination of a plate-girder bridge and a "closed-web" arch: an arch made like a plate girder, but bent. It is shown under construction around 1900, and there is some reason to believe that the location is Bellevue (opposite).

The closed-web arch of 1899 carrying Forbes Street over Junction Hollow.

The bowstring arch of the West End Bridge during a steamboat race, c. 1950.

This structure is a combination of an arch and two cantilevers, carrying Halket Street in Oakland. The photograph was taken in 1893, when the bridge must have been new; it lasted at least until 1921. The railing pattern is identical with that of the Brady Street Bridge.

Suspension bridges are just as simple in principle. This is as it should be, since the *catenaries*, the cables or chains from which the bridge is hung, are inverted arches, arches in tension rather than compression. Most catenaries are made of thousands of wires wrapped together, as in the South Tenth Street Bridge, that are held down at the ends by massive *anchorages*. *Towers* (so called, but in fact bents) support the catenaries so that they have the proper sag to support in a uniform way the loads from the bridge's deck structure that are carried by the *hangers*—usually cables, again—that connect deck and catenaries. The floor structure imparts a stiffening effect that absorbs uneven loads, for example from traffic. In a *self-anchored suspension bridge*, there are no anchorages; rather, compression members in the deck structure counteract the tension in the catenaries.

The only pure suspension bridge in the Pittsburgh area, at South Tenth Street across the Monongahela.

The Durability of Bridges

How long a type of bridge can be expected to last depends on so many circumstances that expectation itself seems almost vain. One has to eliminate from consideration auxiliary or decorative features—reckon without the original railings, lamp-posts, urns, and other things not needed to support the road surface. One has to assume that the main spans continue adequate to handle the loads and volumes of traffic. One has to assume adequate maintenance, or to factor in the amount of inadequacy in a particular case. Then, perhaps, one may make comparisons.

Decay at the Panther Hollow Bridge.

It looks as if the solid masonry arch is the most durable constructional type to be found around here—the viaducts at Washington Boulevard from c.1905, or the bridge of 1889 at Edgeworth over Little Sewickley Creek. Concrete arches have been less durable, perhaps because any exposure of the uncoated reinforcing rods of the past, through cracking of the concrete, led to their rusting and swelling, which forced apart the adjacent concrete, and so on in a vicious cycle. Two bridges carrying Atherton Avenue (Baum Boulevard) over railroads were under construction in 1912, and lasted scarcely six decades. Ditto the Mount Washington Roadway Bridge, built in 1928, down in 1983. The City's John P. Moore Memorial Bridge, built in 1927, fell even sooner, in 1978. Metal trusswork and arches, though requiring paint, seem to do better. The Smithfield Street Bridge's original main spans, built of steel and wrought iron, have been saved by two rebuildings

Down comes the John P. Moore Memorial Bridge in Knoxville, 1978.

Two views of the John P. Moore Memorial Bridge in the days of its hopeful youth: going up in 1927. Concrete is being poured into the formwork, supported by a forest of falsework, for one of the end spans. Formwork is employed to hold poured liquid concrete in shape until it hardens.

The main arch of the John P. Moore Memorial Bridge is still supported by falsework, and concrete is being poured into the formwork of its spandrels.

The Meadow Street Bridge of 1910 after reconstruction of everything above the main arches (above).

The Larimer Avenue Bridge under reconstruction around 1980 (above right).

of the floor structure. The Schenley and Panther Hollow Bridges, steel arches, were built in 1897. Suspension bridges are rarities around here, and our only conventional suspension bridge, that at South Tenth Street, gave considerable maintenance problems some 60 years after its construction.

It looks as if our few masonry arches will last forever if well maintained, and anything dependent on steel—concrete included—only insofar as it can avoid rust.

The enduring nature of stonework despite extraordinary stresses. The Brilliant Cutoff Viaduct, alongside Washington Boulevard.

Some Engineers

John Augustus Roebling (1806–1869)

Though many of the bridges in this area are more or less anonymous products of local public works departments, engineers of national reputation have designed bridges for the Pittsburgh area, and have even begun their careers here.

This was true of the most famous of all American civil engineers, John Augustus Roebling. Having devised a method of manufacturing wire cable, he applied it first to an aqueduct across the Allegheny River for the Pennsylvania Canal—about where the Fort Wayne Bridge is now—in 1844;

then, in 1846, to the second Smithfield Street Bridge; and finally, in 1859, to the second Sixth Street Bridge. By the time the latter was under construction, Roebling's reputation was international, thanks to a suspension railroad bridge he had finished across the Niagara River in 1855. He went on, of course, to design the Brooklyn Bridge, built between 1869 and 1883.

It is strange to consider the cause given for Roebling's coming from Germany to America, a liberalism that brought him into disfavor with the Prussian state, for he was a harsh, rigid family man and employer, who left disagreeable memories. The tale is told, for instance, that after his wife's death he looked for a medium to communicate with her, so that he could utter the words of love he had withheld during her life. When he was convinced that contact had been made, Roebling got her to report in a series of séances on conditions in the next world: and that was all.

Blue collar and white collar at work together on the Larimer Avenue Bridge. August 21, 1911.

Gustav Lindenthal (1850–1935)

Roebling's Smithfield Street Bridge yielded in 1883 to the first element of the present one, the work of Gustav Lindenthal. Born in Brno, Moravia, he came to the United States in the early 1870s, without a formal degree in engineering and construction. Persistence, though, slowly raised his professional status and his competence so that he became, first, an engineer for the eminent Keystone Bridge Company, then for the Atlantic & Great Western Railroad. Finally, at the beginning of the 1880s, he set up his own practice. The Smithfield Street Bridge was his first independent work of any consequence, and brought him national reputation at once. Lindenthal went on to design much more ambitiously, most notably for New York: the Queensboro and Hell Gate Bridges, with Henry Hornbostel as architect. Lindenthal never did span the Hudson, as he had wished, however; that accomplishment was left to his former assistant Othmar Ammann, whose contact with Pittsburgh is limited to the dome of the Civic Arena.

A mock-up of the upper-chord assembly of the Hell Gate Bridge, now at Station Square.

George S. Richardson (1896–1988)

In 1979, George S. Richardson, then the retired founder of Richardson, Gordon & Associates, wrote a memoir of his experiences with the big County bridge-building program of 1924–37. The human element is important, and Richardson and a number of other noted engineers appear in his words.

Richardson came to the County Works Department in 1924 as a draftsman, a post he had been occupying with the American Bridge Company. Vernon R. Covell was County Engineer at the time, and the design of three recent bridges had been done by outsiders. At this time, though, a reorganization was in progress, terminating with a new Department of Public Works for the County, with Norman F. Brown as director. In the new organization the arrangement was more specialized and elaborate than heretofore, with Vernon R. Covell now chief engineer of the Bureau of Bridges and A. D. Nutter, whose name was often to be noted, as chief design engineer. Stanley L. Roush filled the new post of County Architect.

Richardson recalls the new Department as a highly effective one, and one in which he found much interesting work. While it is impossible to attribute a major bridge design, in all its details, to one engineer, Richardson claimed that "typical duties included directing general layout, preliminary design for substructure and supervisor of super-structure design and preparation of contract drawings" for the

McKees Rocks and West End Bridges, and similar work on the George Westinghouse Memorial Bridge. He is credited, too, as principal designer or at least bridge design engineer, of the Liberty Bridge, the second Point Bridge, and the South Tenth Street Bridge, all dating from his time with the County's Department of Public Works.

In 1932 another reorganization took place, with bridge design entrusted to Richardson, who now worked for the County Planning Commission. In 1934 a change occurred again, with a new Allegheny County Authority established to take advantage of Federal Public Works Administration money, with Richardson now "bridge engineer." Under his direction were designs for the Homestead High Level and Jerome Street Bridges, among others. Political difficulties led to his dismissal in 1937, and he soon went into private practice.

Frick and Richardson, an early partnership, was replaced by a practice solely under his own name between 1939 and 1955. In the latter year, Richardson, Gordon & Associates was formed: a consulting firm best known for the Fort Pitt and Fort Duquesne Bridges; other conspicuous works include the I-79 bridge over Neville Island and the present Sewickley Bridge. The firm is now HDR Engineering, Inc.

Early photographs: West End (top) and Westinghouse Bridges (above).

The River Bridges

The Pittsburgh District, U.S. Army Corps of Engineers, publishes navigation charts for the Ohio, Monongahela, and Allegheny Rivers that include their whole lengths in Allegheny County: 15, 35, and 29 miles respectively. Along with other information given, the charts name, locate, and give the profiles of the 48 bridges crossing navigational channels in the county. Further, they give the vertical and horizontal clearances of the navigational channel spans, thus affording a rough idea of their dimensions. The West End Bridge, about a mile down from the Point, is shown with a horizontal clearance of 755.0 feet. This takes into consideration the thickness of the piers that support the arch of the superstructure; the actual distance between the points of support is 788.0 feet. The vertical clearance, for the central part of the span, is 66.3 feet (i.e. 66 and three-tenths feet); this is given in another way too: "elevation of low steel" is 776.3 feet, from which you deduct the 710.0 feet above sea level of water impounded by the Emsworth Dam. Though this is not their purpose, the Navigation Charts make good bridge-recognition books on the rivers.

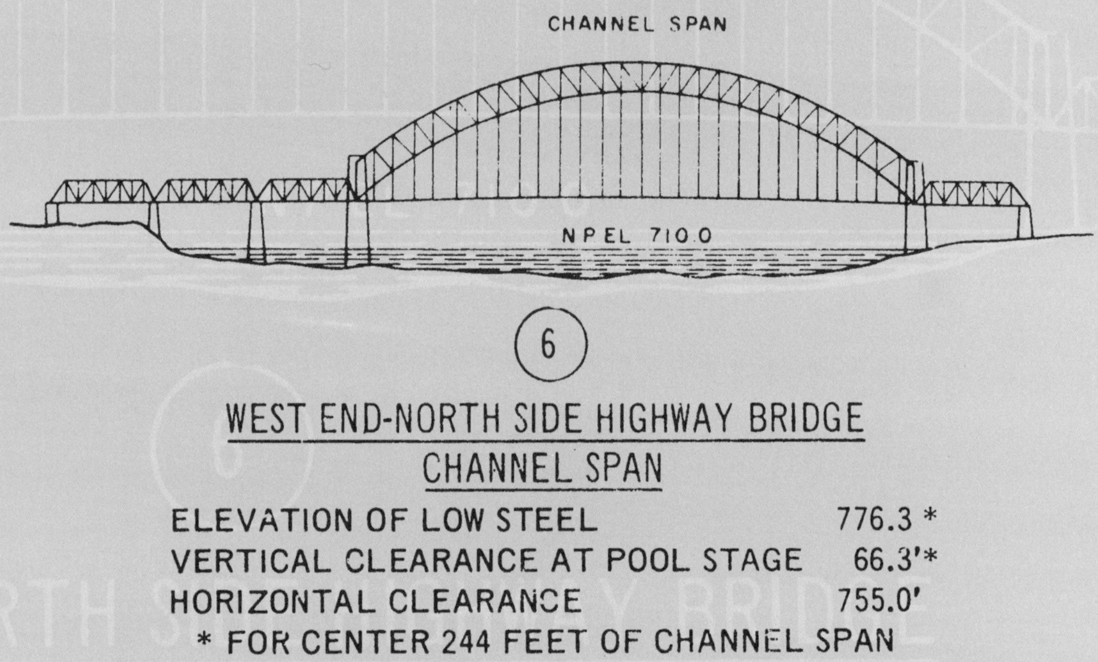

For information on Navigation Charts for the Pittsburgh District, call (412) 395-7502.

You Probably Are Wondering

How many bridges are there in Allegheny County? In 1990 the Southwestern Pennsylvania Regional Planning Commission said 1,674, and has no more recent figure. The City of Pittsburgh *owns* 144 bridges, though seven of those are culverts, presumably of less than eight-foot span and really not counted. Allegheny County, as a corporate entity, maintains 514 bridge structures, but some are listed as culverts and others are as little in overall length as three feet. We can assume that since 1990 a few more real bridges—with main spans of more than eight feet and not merely culverts—have been added to the 1,674 somehow counted back then.

Like a Japanese print: umbrellas on a bridge.

The Photographer's Art

An Album by Clyde Hare

Among the contrivances of humans that seem at first to dwarf their makers, bridges are notable. We swarm over them like two-legged ants on some mighty creation of nature, yet it is we who imagined them and the precise forms of their components, smelted the iron and refined the steel of which they are made, assembled them, and put them to use; we, so small beside them.

A photographer sees a multitude of creative opportunities in the bridges that abound in a place like this: the forms they assume, the ways in which they have occupied the land, and our manners of passing over and under them.

The bridges are but one of the subjects to which Clyde Hare has directed his attention, in a Pittsburgh career that began in 1950. He once recalled, "I made up my mind to take one great photograph each day, one that would live, would be simple, would have something to say."

South Side reminiscence: the Brady Street Bridge looking toward the South Side, early 1950s.

The working river at dusk: downtown from the Eliza Furnaces. May 1962.

The towboat *Homestead* passes beneath the second Point Bridge in 1950.

The last wagon? The Smithfield Street Bridge on a foggy day, downtown in 1954.

The Smithfield Street Bridge before 1985 when the streetcars ceased to cross it.

Walking steel: the Sixty-second Street Bridge, May 1961.

The Pennsylvania Turnpike Bridge over the Allegheny River, 1951: the riveter is on the way to the very end of the truss to pick up fresh rivets.

River reflections on the floor of the Fort Pitt Bridge, June 1984.

The Fort Pitt Bridge: revealing the panorama of town in the early 1980s.

Light, shade, and structure: the old Bloomfield Bridge.

The old Bloomfield Bridge, and the little world below.

Trustworthy old trusses: the meeting of the eyebars.

Limberness and strength: the Smithfield Street Bridge in its old aluminum paint.

Light on the Smithfield Street Bridge: finial, Courthouse tower, and skyscrapers; City of Pittsburgh escutcheon; two grotesques and a lamp bracket before restoration.

Going home: Fort Pitt Bridge approaches.

Crossing Penn Avenue on the pedestrian bridge.

Business, pleasure, and the Fort Pitt Bridge.

Arch, catenaries, and sternwheelers: a Regatta in progress.

Up on the catenaries: Regatta spectators.

The Fort Duquesne Bridge and rowers.

The architectural majesty of the Sixteenth Street Bridge.

Fog-bound birds.

Balloons at daybreak.

The arch in concrete: the George Westinghouse Memorial Bridge.

The arch in steel: the West End Bridge.

Camelback, cantilever, and suspension bridges on the Monongahela.

From Herr's Island at nightfall, on the Allegheny.

Ice on the Ohio at the West End Bridge: January 1994.

Water textures: under the Smithfield Street Bridge.

Fort Pitt Bridge: the day begins.

Ripples on the Ohio: sunrise at the Point, 1994.

Bridges at the Point and their successors, in the 1960s.

Bridges of the Pittsburgh Area:

A Selection from the Present and Past

Of the many bridges in or near Pittsburgh, we have chosen 50 to treat individually, some extant, some gone. We have been inclined to select bridges of elegant or picturesque design, but bridges of raw power in conspicuous places have drawn our attention too. We have attempted variety in structural types and materials, and reasonably good geographical distribution. We begin with the river bridges: first at the Point; up the Monongahela; up the Allegheny; down the Ohio; and then inland. When more than one bridge has preceded an existing one, the present bridge comes first; then we go back in time to the earliest that we show. If the bridge has been demolished, then dates are given in the bridge headline.

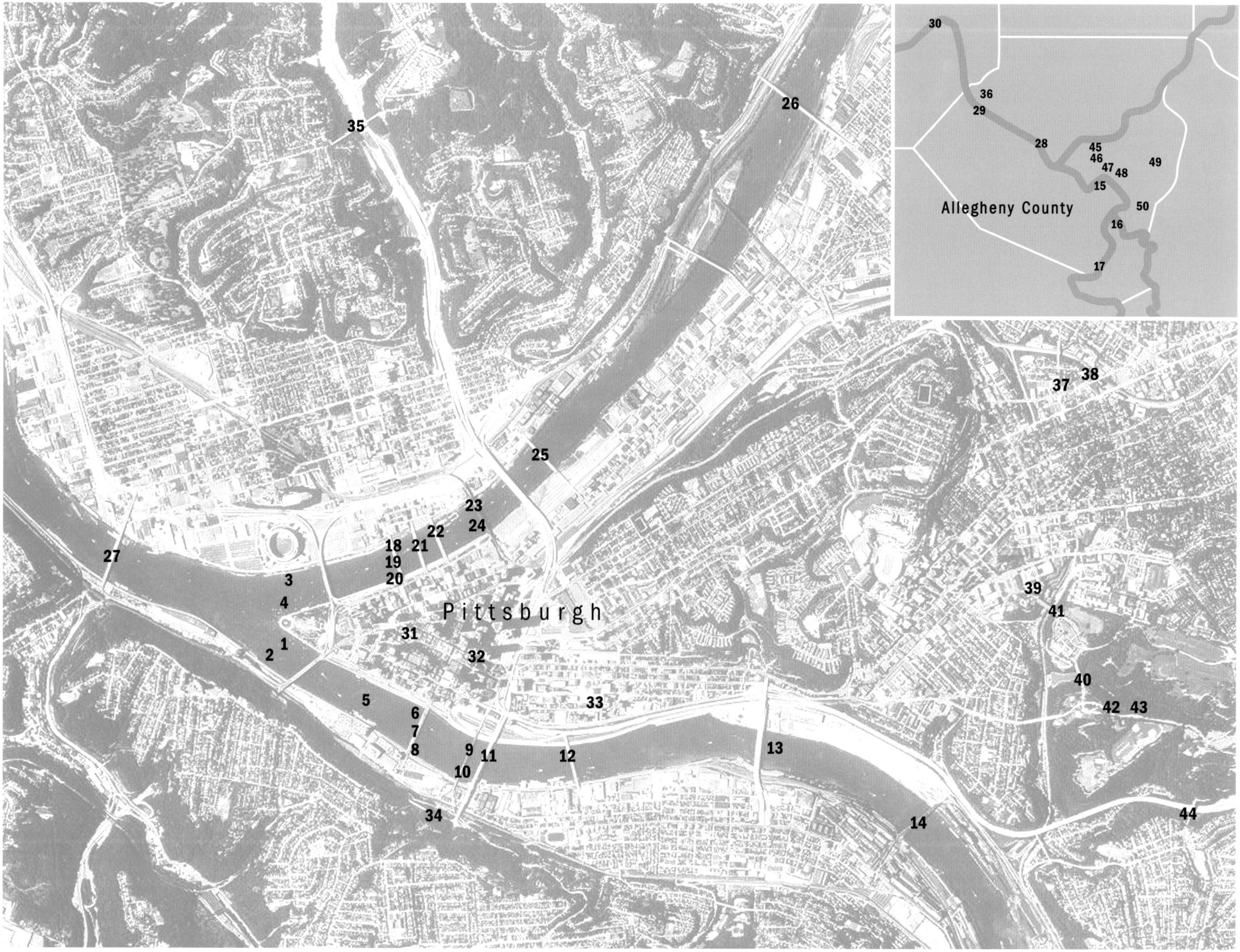
Pittsburgh
Allegheny County
1
2
3
4
5
6
7
8
9
10
11
12
13
14
15
16
17
18
19
20
21
22
23
24
25
26
27
28
29
30
31
32
33
34
35
36
37
38
39
40
41
42
43
44
45
46
47
48
49
50

Legend

If a date is given, then the bridge is no longer extant and the number indicates the site of the former bridge. The bridges are numbered to correspond with the map and according to the sequence in the guide section, following.

Bridges at the Point

1 Point Bridge (1927)
2 Point Bridge (1876)
3 Manchester Bridge (1913)
4 Union Bridge (1874)

Monongahela River Bridges

5 Wabash Bridge (1904)
6 Smithfield Street Bridge
7 Smithfield Street Bridge (1818)
8 Smithfield Street Bridge (1846)
9 Panhandle Bridge
10 Panhandle Bridge (1863)
11 Liberty Bridge
12 South Tenth Street Bridge
13 Brady Street Bridge (1896)
14 Main and Hot Metal Bridges
15 Homestead High Level Bridge
16 Jerome Street Bridge
17 Elizabeth Highway Bridge

Allegheny River Bridges

18 Sixth Street Bridge (Roberto Clemente Bridge)
19 Sixth Street Bridge (1859)
20 Sixth Street Bridge (1892)
21 Seventh Street Bridge
22 Ninth Street Bridge
23 Fort Wayne Bridge
24 Fort Wayne Bridge (1867)
25 Sixteenth Street Bridge
26 Washington Crossing Bridge

Ohio River Bridges

27 West End Bridge
28 McKees Rocks Bridge
29 Sewickley Bridge (1911)
30 Pittsburgh & Lake Erie Railroad Bridge

Inland Bridges

31 Diamond Market (c. 1914)
32 "Bridge of Sighs"
33 Mercy Hospital Footbridge
34 Mount Washington Roadway Bridge (1928)
35 East Street Bridge
36 Sewickley Creek Bridge
37 Atherton Avenue Bridge over the Pittsburgh Junction Railroad (1913)
38 Atherton Avenue Bridge over the Pennsylvania Railroad (1913)
39 Bellefield Bridge (buried)
40 Panther Hollow Bridge
41 Schenley Bridge
42, 43 Tufa Bridges
44 Beechwood Boulevard Bridge
45 Heth's Run Bridge
46 Larimer Avenue Bridge
47 Brilliant Cutoff Viaduct
48 Lincoln Avenue Viaduct
49 Old Stone Bridge, Monroeville
50 George Westinghouse Memorial Bridge

Bridges at the Point

Point Bridge, 1927 – 70

Monongahela River at the Pittsburgh Point

Allegheny County Department of Public Works, engineers;
George S. Richardson, principal designer; Stanley L. Roush, architect

Built by Allegheny County

Under construction, November 15, 1926 (below).

In the foreground, the Monongahela River and the Point Bridge. In the background, the Allegheny River and the Manchester Bridge (opposite).

This bridge had a very unusual form, and, like the Three Sisters bridges on the Allegheny, it assumed this form for art's sake. The City's Art Commission had decided that a curved upper silhouette, like that of the trusses of the adjacent Manchester Bridge, would give an approximately symmetrical look to the Point.

According to George S. Richardson of the County Department of Public Works, the design of the new Point Bridge was entrusted to him in 1924, though he attributed the basic form to a designer named Marks in the office of Stanley L. Roush, the County Architect. The bridge was framed as a cantilever bridge apparently similar in structure to the Brady Street Bridge. The main span was

Waiting for the end.

670 feet. The massive-looking staple-shaped portals were largely visual devices, connected to the cantilevers only at four points, so as to tie the cantilevers to anchorages.

As the time of demolition became foreseeable, a movement to re-use the Point and Manchester Bridges gained intensity—the idea was to build upon them in the medieval manner still visible in the Ponte Vecchio in Florence—but to no avail, and the Point Bridge came down late in 1970.

The end.

Point Bridge, 1876 – 1927

Edward Hemberle, engineer

Privately built

This was a rigid suspension bridge of a type apparently unique in the United States. The paired towers, 110 feet high and braced halfway up by portal arches, were quartets of hollow wrought-iron posts. From these hung wrought-iron eyebar catenaries that were braced from above with pairs of Howe trusses whose upper chords were straight eyebar chains meeting at the center of the 800-foot main span. The 145-foot approach spans were deck Pratt trusses, unsuspended.

The structure was never satisfactory, and its floor had to be greatly repaired in 1903. In 1927 it came down, and the second Point Bridge, just built a few feet upstream, took over.

The bridge in 1889. The Duquesne Incline is in the background.

The two Point Bridges side by side, in 1927.

Manchester Bridge, 1913 – 70

Allegheny River at the Pittsburgh Point

Pittsburgh Department of Public Works; T. J. Wilkerson, engineer

Built by the City of Pittsburgh

A scene of 1916: the first Point Bridge, the Manchester Bridge, a roller-coaster, and the Exposition Buildings.

In 1907 the Union Bridge, a covered wooden toll bridge of 1874, was demolished at the demand of the U.S. War Department because its low clearance, about 30 feet, inhibited navigation. The new bridge, begun in 1911 and officially called the North Side Point Bridge, had two truss spans 531 feet long, giving a 70-foot vertical clearance above the 703-foot river elevation of the time (present elevation is 710 feet). The bridge was in the main an artless structure, but the two portals were ornamented, in 1915, by large bronze sculptures by Charles Keck, a New Yorker who had been collaborating with the architect Henry Hornbostel on several major Pittsburgh projects. Represented were two of George Washington's companions in exploring this area, Christopher Gist and Guyasuta, with a legendary steelworker and coal miner on the other portal.

The Pittsburgh Renaissance and the development of Point State Park foreshadowed the demolition of the bridge, and it fell to explosives in 1970. The portal sculptures now stand outside the Pittsburgh Children's Museum on the North Side; salvaged too were medallions, lion's heads, torches, eagles, and shields, which had contributed more modestly to alleviate the bridge's plainness.

A falsework collapse during construction in 1912.

The Manchester Bridge in its last days, with ground cleared for Point State Park.

The Manchester Bridge with Charles Keck's portal sculptures, during the late 1950s.

Demolition under way.

The blast that felled the northern main span.

Salvage from the portals: Christopher Gist, frontiersman and companion of George Washington; Guyasuta, leader of the Senecas; Joe Magarac, legendary steelworker; and Jan Volkanik, hero of the coal miners.

No. 68.
U.S. ENGINEER OFFICE,
PITTSBURG PA.

Union Bridge, 1874 – 1907

Allegheny River at the Pittsburgh Point

Privately built

We are so used to thinking of Pittsburgh as the Iron City, or later the Steel City, that it is hard to believe that in 1890 the Union, Ninth Street, Sixteenth Street, and Forty-third Street Bridges on the Allegheny River were wooden covered bridges, the last of which lasted into the early 1920s, and that the South Tenth Street Bridge of 1861 on the Monongahela was one also. All these had been built privately as toll bridges, and the old arch-assisted Burr truss spans of wood, though they demanded numerous stone piers in the rivers, seem to have needed less capital investment than larger spans of iron.

The Union Bridge and the first Point Bridge, c. 1900 (opposite).

The exposed state of the bridge near the end of its existence.

A staged confrontation in 1903 by the *Mayflower*, a packet. The bridge's clearance of about 30 feet was inadequate, even with the stacks and pilot house roof lowered. A banner on the boat reads, "Here I is, can't go. Low bridges, gee wiz."

Monongahela River Bridges

Wabash Bridge, 1904 – 48

Monongahela River at Mile 0.5

Boller & Hodge, engineers

Built by the Wabash-Pittsburg Terminal Railway

The Wabash-Pittsburg Terminal Railway, a suicidal enterprise that began operations in 1904 and was in receivership in 1908, was the brain child of George Jay Gould, son of the notorious Jay Gould, that supreme financial schemer of the Victorian period. The Wabash was intended to challenge the near-monopoly of the Pennsylvania Railroad on freight in the Pittsburgh area. From Jewett, Ohio, where a railroad passed that was already in Gould's hands, it was only 60 miles to Pittsburgh: but 60 miles of daunting terrain, requiring bridges, tunnels, and embankments, some of them very long. In Allegheny County, where some of the severest terrain lay, were the Greentree Tunnel, 4,718 feet long, and the Mount Washington Tunnel, 3,342 feet long, both threatening engine-crew suffocation. The Mount Washington Tunnel led almost directly to a 1,504-foot cantilever bridge across the Monongahela; the piers that supported the 812-foot main span of this bridge are still in

A collapse during construction in 1903.

The Wabash Bridge around 1906.

The packet *Virginia*, to become famous for washing up in a cornfield in 1910, turns beneath the Wabash Bridge. Note the lowered smokestacks.

Wharfboats and the Wabash Bridge. The presence of the bridge and the towboat *Rival* beneath it places this scene between 1904 and 1928.

place, one at the Gateway Clipper landing, one near the end of Stanwix Street. Beyond, at Liberty Avenue, lay the Wabash Terminal itself, with a six-track balloon trainshed 450 feet long, and beyond this a palatial passenger station and such storage tracks as could be accommodated 70 feet above street level.

The trouble was that everything in the last five miles was a bottleneck. The Rook Yard, about five miles from the Terminal in Pittsburgh, was the first place where locomotives could uncouple and turn around. In the Terminal trainshed and the exiguous yard trackage alongside, locomotives were trapped, steaming and smoking, until their trains were ready to back out of town.

The receivership in 1908 led to absorption by the new Pittsburgh & West Virginia Railway in 1917. The Terminal trainshed was wrecked by fire in 1946, the bridge came down in 1948, and the palatial terminal building on Liberty Avenue in 1955.

The "Seldom Seen" arch was built in 1902 through an approach embankment on the south side of Mount Washington. It is faced on both sides in rock-faced sandstone of regular cut, and inside is composed of 21 arches of brick that overlap to compose a skew vault, that is, one whose axis is not at right angles to its exterior. The vault is about 45 feet in span and about 92 feet deep. It is south of and within sight of Saw Mill Run Boulevard, and east of Woodruff Street, which descends the rear of Mount Washington.

The Seldom Seen arch over Saw Mill Run.

A slightly synthetic restoration of colors has the superstructure of the Smithfield Street Bridge a deep gray-blue and the floor structure cinnamon, as in 1883, and the portals buttermilk, as seems to have been the case in 1915.

Smithfield Street Bridge

Monongahela River at Smithfield Street, below Mile 1

Built 1881–83; various additions and alterations
Gustav Lindenthal, engineer of original part

Built by the Monongahela Bridge Company
Present owner: PennDOT

The portal after the 1891 expansion (to the left).

A lenticular, or Pauli, truss operates on a very simple principle: its upper chord is an arch, which thrusts outward; its lower chord is a catenary, like the draped cable of a suspension bridge, which pulls inward. The horizontal forces cancel each other out, and vertical forces are transmitted to posts and hangers, and counters within the truss absorb dynamic and off-center loads. In profile, the truss suggests a double-convex lens seen on edge: hence, "lenticular."

The present Smithfield Street Bridge has two lenticular main spans of 360 feet each, the largest ever built in the United States; they are smaller, say, than those of the Royal Albert Bridge in Cornwall—a pair of 455-foot spans—but they are more graceful

The bridge in the 1890s, looking downtown (right). The B&O Station appears to the right. The turnbuckle detailing of the trusses then was different at the junctions of the diagonals, though these are probably no more than decorative covers.

and delicate looking seen from the side. It is curious to note, by the way, that the arches of the Schenley and Panther Hollow Bridges are also 360 feet in span, though, seen from the ravines they cross, the park bridges seem much greater.

The downstream half of the Smithfield Street Bridge is the original, and KLOMAN PATENT PROCESS 1881 is stamped into the eyebars that form its catenaries. Andrew Kloman was a rival of Andrew Carnegie and proprietor of the Pittsburg Bessemer Steel Company in Munhall, but it was in fact not Bessemer steel but the superior open-hearth steel that was forged by Kloman's process into the eyebars and used as well for the upper, arch-like chords, and the portal posts, diagonals, and connecting pins, all else in the trusses being of the more customary wrought iron. The roadway and sidewalks were originally of gum wood and white pine. The bridge was widened upstream in 1890–91, then widened still further in 1911 to accommodate two trolley tracks. At this point the portals of 1883, faced in ponderous compositions of cast iron, and the annexed portals of 1891 were removed and the entrances to the main spans left skeletal. But around 1915 the City Architect, Stanley L. Roush, designed the present light-hearted portals, more or less mock Gothic and including grotesques of workmen at the springings of the arches.

In 1934 the old wrought-iron floor members were replaced by ones of aluminum, greatly lightening the bridge's dead weight, and the rather fussy iron railings were replaced by plain hollow aluminum railings. In 1994 the aluminum floor structure was replaced by a new one of steel, and the bridge was re-lit and repainted in ways representing a compromise with historical accuracy: trusses a deep gray-blue, floor structure cinnamon, and the portals buttermilk as they appear to have been in 1915.

In 1883, the bridge also had three wrought-iron plate-girder spans north of the main spans and four to their south; there have been alterations since.

The bridge has a multitude of distinctions: National Historic Landmark, National Historic Civil Engineering Landmark, and City Historic Structure.

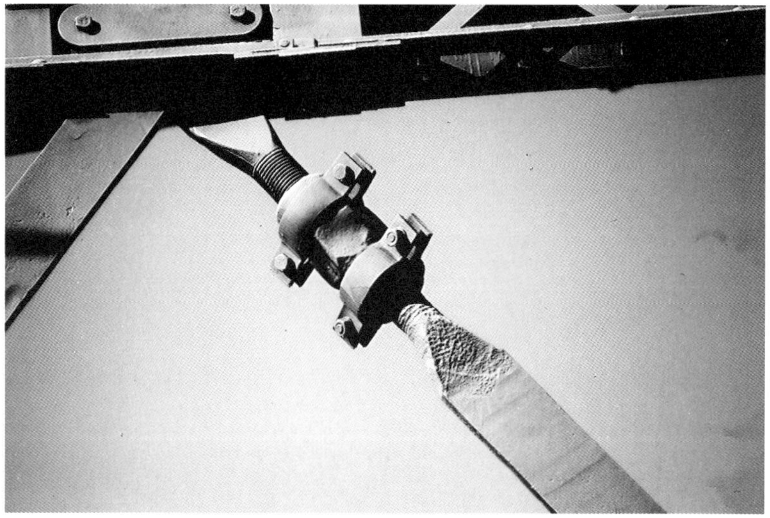

A turnbuckle adjusts the tension on a minor truss member (above).

The catenaries of the trusses are formed of pin connected eyebars (below).

Looking south to the Pittsburgh & Lake Erie passenger station, now the Landmarks Building at Station Square, and the trolley shelter added in 1912. Both walks and roadway are decked in wood (opposite).

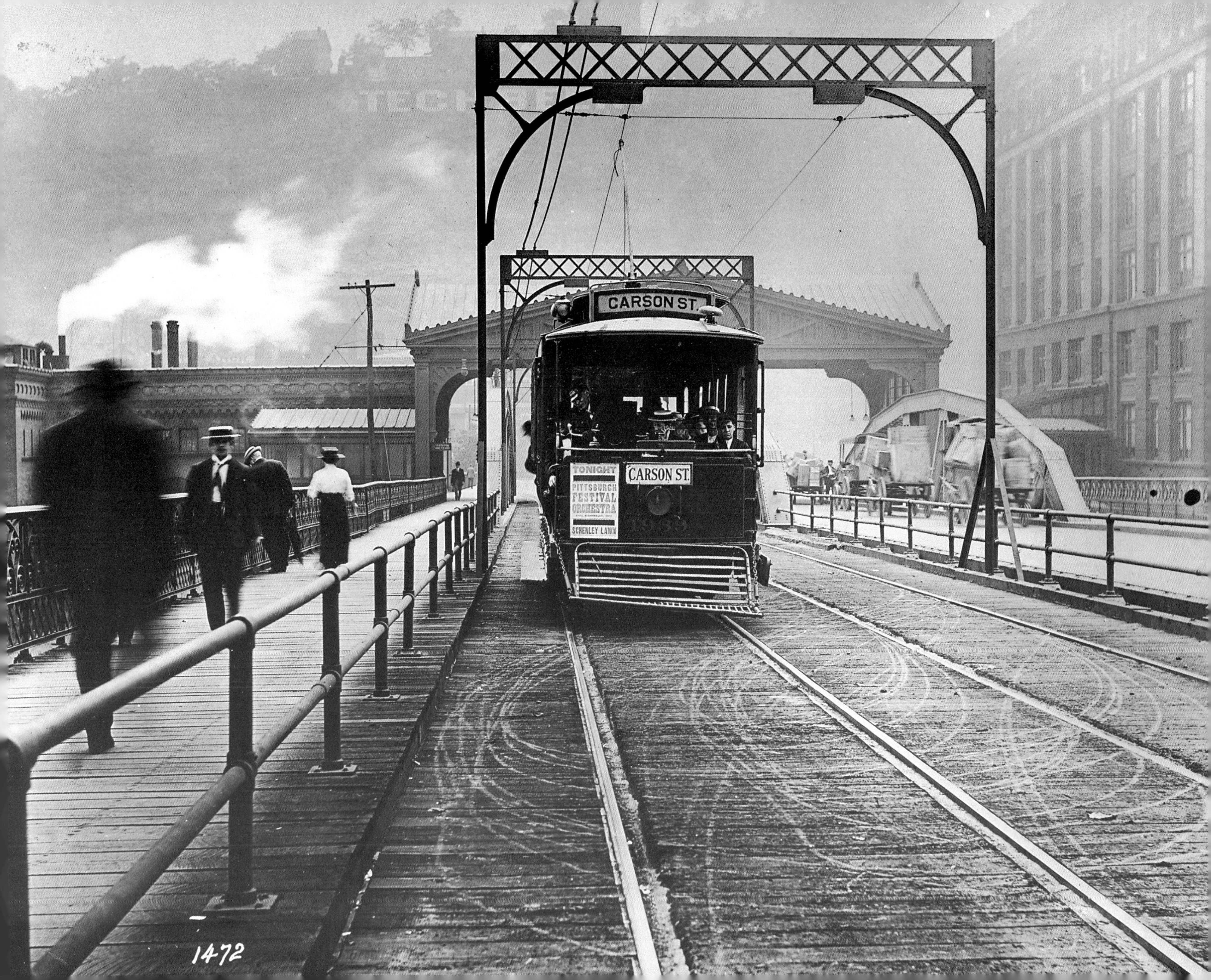
CARSON ST.
TONIGHT
PITTSBURGH
FESTIVAL
ORCHESTRA
SCHENLEY LAWN
CARSON ST.
1472

Smithfield Street Bridge, 1818 – 45

Lewis Wernwag, designer

Built by the Monongahela Bridge Company

Russell Smith's picture after the freshet of 1832. The building at the right appears to contain the toll booths at its sides.

The first Smithfield Street Bridge, finished in 1818, was very likely the first bridge, anywhere, across any of the three rivers that flow through Pittsburgh. As early as 1810 there was a plan to build bridges of a chain suspension type, one across the Monongahela, one across the Allegheny, but the War of 1812 appears to have aborted the project. The bridge actually built at Smithfield Street, starting in 1816 and finished in 1818, had abutments and seven piers of stone, with eight covered Burr trusses—wooden trusses reinforced by wooden arches. The designer was Lewis Wernwag of Philadelphia, whose "Colossus," spanning the Schuylkill River there, had a single wooden trussed arch of 340-foot span. This span was a record-breaker for an arch in any material, but in Pittsburgh the 1,500-foot length had 188-foot spans. In 1832 a "freshet" took out the first two spans at the Pittsburgh end; they were replaced, but the repaired bridge went in a matter of ten minutes in the Great Fire of April 10, 1845.

Smithfield Street Bridge, 1846 – 83

John Augustus Roebling, engineer

Privately built

On the piers of the earlier bridge, in 1845–46, went a wire-cable suspension bridge by the German-born John Augustus Roebling, who had already established the manufacture of wire cables in the United States and who had just completed a cable-suspended aqueduct for the Pennsylvania Canal between Allegheny and Pittsburgh. His masterpiece, of course, was to be the Brooklyn Bridge, begun in 1869 and with a main span of 1,596 feet as opposed to the 188-foot spans at Smithfield Street. The new superstructure over the Monongahela took seven months to build. The cast-iron towers held pendulums from which the actual bridge cables were suspended, so that large dynamic loads could be absorbed without shock. The bridge did indeed have to bear heavy industrial traffic, and in time became overstrained. It came down, piers and all, in 1883 when its successor was opened.

The bridge around 1880. The Pittsburgh & Lake Erie Railroad, which began operations in 1879, seems to be busy. To the far right is the Clinton Furnace, the oldest blast furnace to operate successfully within the present Pittsburgh city limits.

Panhandle Bridge

Monongahela River at Mile 1

1903

Built by the Pittsburgh, Cincinnati & St. Louis Railway
Present owner: Port Authority of Allegheny County

This bridge now serves the Port Authority's Light Rail line into town, lying just downstream from the Liberty Bridge. Its single Pennsylvania truss of about 360 feet, its two "camelback" Pratt trusses, and a pair of deck Pratt trusses form a total length of 1,172 feet. The bridge replaced an original of 1863, carrying the present line to an underground route and over places meaningful in Pittsburgh and American history. Opened in 1834, the original tunnel under Grant's Hill extended the Pennsylvania Canal to a projected connection with the Chesapeake & Ohio Canal. A realignment of the tunnel route brought the Pittsburgh, Cincinnati & St. Louis Railway (the Panhandle Line, a Pennsylvania subsidiary) to and across the Monongahela from Union Station. Further realignment and extension created the present Port Authority subway route in town. The route passes over or near the places where the *New Orleans* initiated steam navigation on the Western Rivers; where the architect and engineer Benjamin Latrobe built a shipyard for Robert Fulton in 1813–14; and where the first emergency application of a railroad air brake was made.

A modern view with the bridge serving the Light Rail; the ramp to the Arlington Avenue line is in the background to the left (right).

The main spans, with the Bluff in the background (below).

Panhandle Bridge, 1863 – 1903

Original owner: Pittsburgh, Cincinnati & St. Louis Railway

The Panhandle Bridge of 1863 was of interest to *Engineering Record* at the time of its replacement in 1903, being of a bygone technology:

The old bridge consisted of one 260-foot double-track through span about 40 feet deep and four 186-foot deck spans. The trusses were of Linville type with very short panels and double-intersection diagonals. All compression members were made of cast iron, and the details throughout were of the earliest types of long-span iron railroad bridges in the United States. The top chords and end posts were made with double tubular cross sections and the end posts were trussed... with pairs of screw-ended rods which had center bearings on saddle castings and virtually hog-chained the members. The top lateral and sway-brace struts were double flanged castings with open panels. The portal girders resembled modern portals, but were made with girders, brackets, and web trussing of cast iron. The vertical posts were made with twin uprights connected by webs having large open panels....

The 1863 bridge, as shown in White and von Bernewitz, *The Bridges of Pittsburgh.*

The Panhandle Bridge as it was in 1893.

The floor-beams were made with pairs of channels king-post-trussed with pin-connected rods and seated in chairs resting on the bottom chords. The track was carried on wooden stringers. The main diagonal rods were square bars with screw ends and nuts bearing on bevelled seats cast on the top chords and on the feet of the vertical posts.

The old bridge was built by the Keystone Bridge Company nearly forty years ago, had sustained constant and increasing service ever since, and was found in a remarkably good condition when taken down.

By the time Joseph White and M. W. von Bernewitz quoted this article in *The Bridges of Pittsburgh* it was 1928, and all the more quaint were all the cast iron and trussed-together floor members.

A view seemingly of the 1870s shows this main span approached by covered deck spans that appear to be of wood. The danger of rot must have been great, and the 1893 view shows Whipple trusses of iron or steel.

Liberty Bridge

Monongahela River above Mile 1

1928
Allegheny County Department of Public Works, engineers; George S. Richardson, principal designer

Built by Allegheny County
Present owner: PennDOT

The opening of the Liberty Tubes in 1924 seemed to offer a quick penetration of Mount Washington and speedy distribution of automobile traffic to points of the South Hills too far for a walk to the trolley: no straphanging, no jamming-together of houses. But bad ventilation led one day to a panic, and that to a closing until a ventilation system could be contrived. Even after that was done, a bridge envisioned as early as 1910 remained incomplete until 1928. The Liberty Bridge, once opened, linked the Tubes with the Boulevard of the Allies as it descended toward town.

The suspended span is shown under construction (above).

The completed bridge before the 1983 reconstruction (right).

A view from 1930 shows the pier towers at their full height (following page).

STANDARD SUPPLY & EQUIPMENT CO
MINE AND INDUSTRIAL SUPPLIES
CONTRACTORS EQUIPMENT

The traffic circle in front of the Liberty Tubes in its original form, suggesting that the pressure and velocity of traffic were not major problems.

Two views from 1960 or so, with the railings and lamps of the Liberty Bridge itself intact, but with the circular island much reduced.

The architectural character of the Liberty Bridge was greatly impaired in 1983.

The bridge had two 420-foot main spans with lower chords shaped like segmental arches, though in fact they consisted of two cantilever trusses and a suspended span. In the 1920s a continuous girder over more than two points of support presented too many variables to be calculated, and the continuity had to be broken. The suspended span, artfully concealed, is at the center of the northern main span.

As built, the piers were carried up to deck level as rusticated concrete towers, and a malleable-iron railing and posts, delicately and Classically patterned, ran along the sidewalks. Remodeling of the bridge in 1983, involving widening of the roadway, removed all the lamps and railings and the decorative tops of the piers, so that the bridge now offers nothing but the view to the crosser's eye.

No. 12252, So. 10th St. Bridge,- Wrapping Cable, 5-3-32.
Dabbs & Bayer

South Tenth Street Bridge

Monongahela River at Mile 1.4

1933
George S. Richardson, with A. D. Nutter and Sydney A. Shubin, engineers

Built and owned by Allegheny County

This is the only conventional suspension bridge in Allegheny County, the Three Sisters on the Allegheny River being, in fact, quite unusual. "Conventional," in that the catenaries are wire cables held down at their ends by masses of concrete called anchorages.
As suspension bridges go, the South Tenth Street Bridge is not particularly large: 725-foot main span, 1,275-foot overall length.

The bridge has had a multitude of maintenance and repair problems, including broken wires in the catenaries, but repairs in 1994 took care of the major troubles.

The only conventional suspension bridge in Pittsburgh (right). The dinosaurs on the top of the tower are not official.

In the foreground is a curved casting that bends the catenary cables down toward an anchorage (opposite).

Four hangers and a stiffening girder on the South Tenth Street Bridge (above).

Stiffening girders in place: April 22, 1932 (right).

Brady Street Bridge, 1896 – 1978

Monongahela River near Mile 2.5, between Brady and South Twenty-second Streets, Pittsburgh
Built by the City of Pittsburgh

The Brady (or South Twenty-second) Street Bridge stood a very short distance upstream from its replacement: the Birmingham Bridge. Built by the City, it was the first toll-free river bridge in Pittsburgh, an impressive visually (though not structurally) unified trio of trusses plus approaches for a total of 2,250 feet in length with a 520-foot central span and 260-foot end spans. The silhouette, later to be found in the second Point Bridge, is rather dinosaur-like: a single upper chord curving down toward the portals and an arch with hangers over the main span.

The bridge functioned humbly and more or less satisfactorily, though pier reconstruction was needed in 1909 and demolition was scheduled for 1969. Concrete bents for the Birmingham Bridge were in place in 1970, but in fact

1979-22nd.St.Bridge from Blvd.of Allies looking South-10/27/36 Carter

the Brady Street Bridge was not blasted down until the late spring of 1978. The contractors in charge of the demolition bungled, blocking the Monongahela navigational channel for 12 days.

Railings from the Brady Street Bridge, fancy linear panels of Late Victorian wrought iron, can be seen in front of the lower station of the Monongahela Incline.

During the repairs of 1909. As is evident, part of the top chord is a dummy, added to maintain the continuity of the curve (above).

The "basket-handle" portal arch seems to be curved for decorative reasons (left).

The Brady Street and Birmingham Bridges: a contrast in technologies, visible in 1978 (above). **Just after the demolition blasts** (opposite).

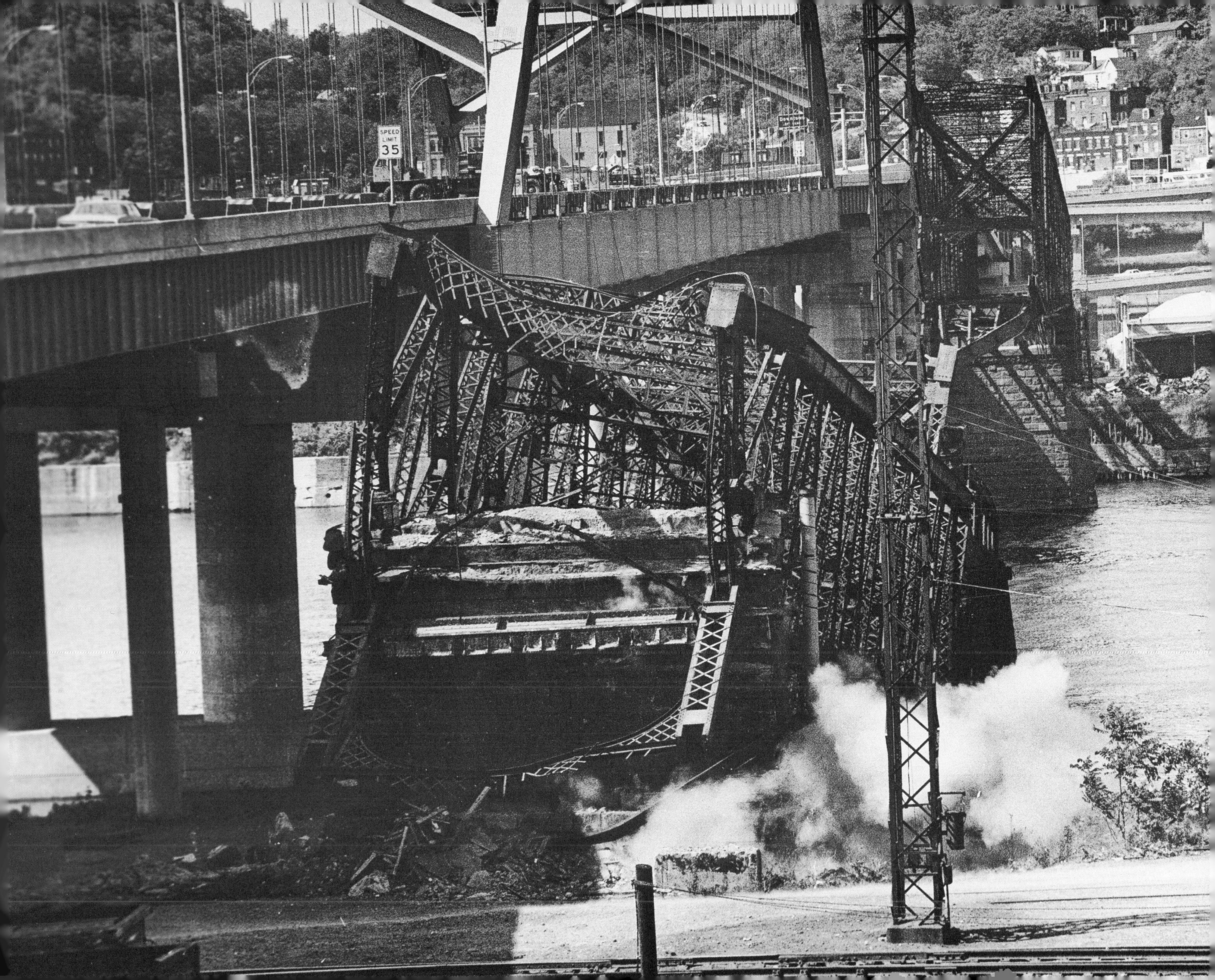
SPEED
LIMIT
35

Main and Hot Metal Bridges, Monongahela Connecting Railroad

Monongahela River at Mile 3

1903; numerous changes
Structural Department, Jones & Laughlin Ltd., engineers

Built by the Monongahela Connecting Railroad
Present owner: Urban Redevelopment Authority, City of Pittsburgh

Hot metal is iron just smelted in a blast furnace, its slag poured off, its temperature still close to 3,000° Fahrenheit. Unless it is to be cast into pig—small ingots—for remelting at a later date, it is obviously economical to keep this new iron in a molten state during its conversion into wrought iron or steel. Yet the Pittsburgh Works of the Jones & Laughlin Steel Corporation, substantially closed down in 1987, began during and after 1852 as a group of small independent companies acting in concert, with the Eliza and Soho blast furnaces on the right bank of the Monongahela and all other operations on the left. This necessitated the casting of hot metal into pig, ferrying this across the river, and wastefully reheating. In 1882 a bridge was built to carry pig across by rail: later, presumably when a fully coordinated metalworking process was evolved in a united company, hot metal crossed directly from the furnaces. In 1928, the six Eliza Furnaces were sending across 4,300 tons

The two bridges stand side by side: the so-called Main Bridge (left in the bottom picture) and the supplementary Hot Metal Bridge (right in the bottom picture).

of hot metal a day, confident that the Bessemer converters and open-hearth furnaces of the South Side Plant stood ready to receive it.

The so-called Main Bridge served the general purposes of the Monongahela Connecting Railroad. It is 1,054 feet long with six spans, the main one being 326 feet long.

A few feet downstream is the Hot Metal Bridge, on which the ladle cars crossed. Its single track was lined with heavy steel, for a spill of hot metal into the river would have had sensational results.

The bridges are destined to be an important traffic link between the cleared land of the J&L sites as they develop.

The bridges in the summer of 1998. The portal structure in the foreground is now demolished. The Main Bridge (left) will carry road vehicles, and the Hot Metal Bridge (right) will carry bicycles and pedestrians.

Homestead High Level Bridge

Monongahela River above Mile 7, between Homestead and Pittsburgh

1935–37
Allegheny County Authority, engineers; George S. Richardson, chief engineer

Built and owned by Allegheny County

By the early 1930s, "Brown's Bridge," a multi-span bridge at Homestead close to water level, was seen as too narrow and a hazard, approached over tortuous roads and railroad grade crossings. The newly constituted Allegheny County Authority, operating with funding from the Federal Public Works Administration, designed a new and higher-set bridge that overrode the obstacles in a cantilever configuration that put one, not four, piers in the river.

Brown's Bridge hobbles across the Monongahela, which the High Level Bridge takes in two bounds.

The Homestead High Level Bridge carries traffic without effort over "the Ward," later replaced by an extension of the USS Homestead Works.

A steam sternwheeler pushes a tow of barges, late 1930s.

The river spans of the Homestead High Level Bridge with the approach spans to the right (above).

The secret of the Wichert truss is this diamond-shaped member, which distributes the forces of the adjacent spans (right).

In that pre-computer time, engineers avoided the continuous trusses that would have offered notable practical advantages *if* they had been able to figure out, under all circumstances, how the members were being loaded. However, shortly before, Ernest M. Wichert, an engineer practicing near Pittsburgh, had invented a four-sided hinged assembly to rest on each pier, allowing the adjoining trusswork of the spans freedom of movement, thus distributing stresses within the whole length of the bridge. Wichert's royalty for this project was $25,000.

There are two arch-like central spans of 533 feet and two of 291 feet. The entire length is 1,654 feet.

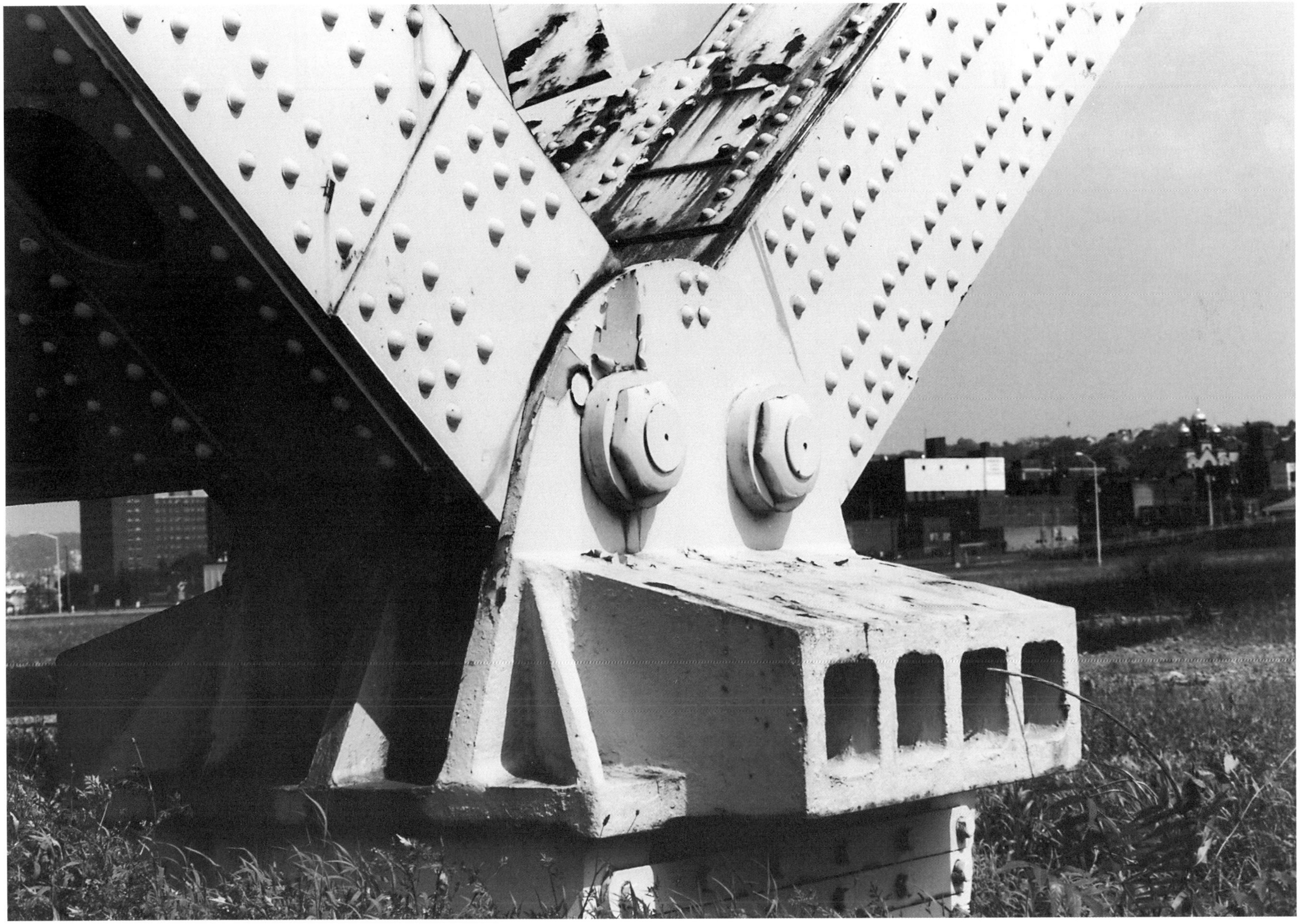

A hinge.

Jerome Street Bridge

Youghiogheny River at McKeesport

1937
George S. Richardson, engineer

Built for Allegheny County
Present owner: PennDOT

Great forces converging on a single point.

In 1932, the Allegheny County Department of Public Works was replaced by the Allegheny County Authority, structured to receive money for approved construction projects from the Federal Public Works Administration. One such project was the Jerome Street Bridge, a crescent arch of 315-foot span about a quarter-mile from the Monongahela.

The bridge is now named after George H. Lysle, the vehemently anti-union mayor of McKeesport between 1913 and 1941.

The graceful crescent arch is almost unique in this area; two others are part of the McKees Rocks Bridge.

Elizabeth Highway Bridge

Monongahela River at Mile 23, between Elizabeth and West Elizabeth

1951

Built and owned by PennDOT

Here is a not-uncommon modern type, a "continuous cantilever," the main span being 400 feet with 210-foot arms. The bridge has long approaches that create a multitude of spans at both ends.

There are 30 spans in all.

The Elizabeth Highway Bridge of 1951 is the newest one shown in this selection of bridges.

Under the Seventh Street Bridge, with the North Shore beyond.

Allegheny River Bridges

The "Three Sisters" Bridges: Sixth Street, Seventh Street, and Ninth Street

Allegheny River at approximately Mile 0.6

Opened between 1926 and 1928

Allegheny County Department of Public Works, engineers; T. J. Wilkerson(?), designer; Stanley L. Roush, architect

Built and owned by Allegheny County

All three of these bridges replaced earlier ones, the Sixth Street Bridge being the fourth on the site since 1819. Our photograph of the confrontation between the towboat *Duquesne* and the third Sixth Street Bridge (see page 4) illustrates why three closely spaced bridges, originating at different times and having different forms of construction, were replaced to a uniform design. The old and low Union covered bridge at the Point had been demolished in 1907, yielding in 1913 to the Manchester Bridge that had a much greater vertical clearance. In 1918, the U. S. Secretary of War ordered the Fort Wayne Bridge, just upstream, raised so as to give a 48-foot vertical clearance. A new policy of encouraging commercial navigation on the Allegheny was in force, and bridge construction had to offer clearances for pilot houses and smokestacks. The Sixth, Seventh, and Ninth Street Bridges, whose bonds were approved by the Allegheny

The Sixth Street Bridge was renamed the Roberto Clemente Bridge in 1999.

Constructing the Sixth Street Bridge: June 15, 1928.

January 25, 1927: The Seventh and Ninth Street Bridges, the later with temporary cantilever struts. The ends of the stiffening girders are marked by obelisks, now gone.

County voters in 1924, were to have a 47-foot vertical clearance. Further red tape was requirement for approval by Pittsburgh's Art Commission, which had some authority over the appearance of bridges in the city. Here the commission, under gentle prompting of the County Architect Stanley L. Roush, chose a graceful suspension form rather than the more ungainly truss forms available.

The result was identical—to the eye—in each case: a self-anchored suspension bridge, of a type unbuilt thus far in the United States, with a main span of about 440 feet, side spans half that length, and catenaries formed of pin-connected eyebars that, instead of having their ends held *down* by anchorages, had them held *out* by a stiffening girder, seven feet deep, that ran along the bridge deck, half above and half below to dissemble its thickness. The sag of the catenary, one quarter of its length in the main span, the very considerable rise of the deck, and a rather rigid system of eyebar hangers from catenary to deck created a structure of upright and inverted arched forms.

In a conventional suspension bridge, the catenaries are built-up affairs of wires upon wires, and indeed a single wire, flown across a gorge on a kite, was the start of a catenary of the Niagara railroad bridge attempted in 1847. Here, where the catenaries were chains of eyebars, every link had to be in place and connected to the longitudinal girders and the hangers—the bearing structure had to be complete, that is—before it was useful at all. In fact, the bridges were fitted with diagonal struts and built as cantilevers from the towers toward closure in the middle. Though the "Three Sisters" seem like models of functional beauty—the American Institute of Steel Construction, indeed, gave the Sixth Street Bridge a beauty prize in 1928—the original choice of the suspension form led to a fussy solution. Yet let us be just: the falsework needed to build a truss or an arch was much more extravagant of temporary material in many cases.

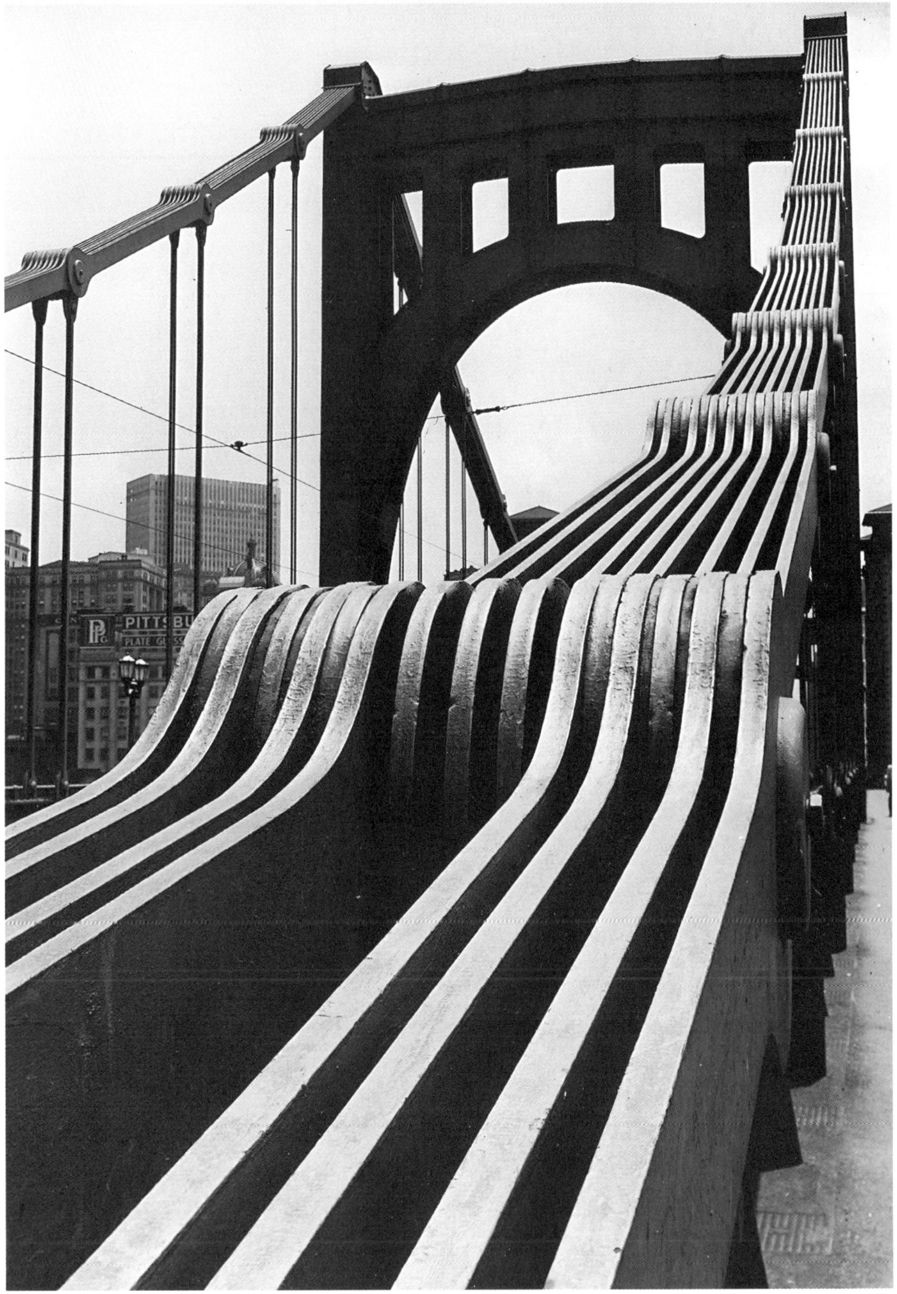

The pin-connected eyebars of the catenaries and the eyebar hangers make an unusually rigid structure.

Sixth Street Bridge, 1859 – 92

John Augustus Roebling, engineer

Privately built

The first bridge on this site, the St. Clair Street Bridge, dated from 1819 and was a wooden covered bridge of six spans, probably Burr trusses with wooden arches built into them, and a total length of 1,017 feet. The designer was a Mr. Lothrop, the contractor.

In the late 1850s this came down, to be replaced by John Roebling's third and last executed bridge in Pittsburgh. This had two main spans of 344 feet, with shore spans of 177 and 171 feet. The floors were suspended from wire hangers that were suspended from wire catenaries. The towers were quartets of cast-iron columns about 27 feet above deck level, not including their ornamental tops. A fire in 1881 resulted in the replacement of some wooden floor members by wrought-iron ones. The advent of the electric streetcar, which imposed heavy moving loads, and the increase of traffic demanded a wider and more rigid bridge, and Roebling's bridge came down in 1892.

The second Sixth Street Bridge, intact (above) and under demolition (below), seen from the shipyard of James Rees & Sons Company, near the Point.

Roebling's Sixth Street Bridge in the 1880s. Upstream is Gustav Lindenthal's Seventh Street Bridge of 1884.

The third bridge on this site, designed by Theodore Cooper, on a smoggy day in 1911.

Sixth Street Bridge, 1892 – 1927

Theodore Cooper, engineer

Privately built

The third Sixth Street Bridge was the work of Theodore Cooper, designing for the Union Bridge Company. The main spans were 440 feet long, each having through trusses of the camel-back type with upward-angled upper chords, and were twice as wide as the previous bridge. Replacing the one with the other with the minimum of disruption of traffic was a remarkably tricky process, described more or less clearly by White and von Bernewitz.

Come 1927, the third Sixth Street Bridge entered a new phase of its career. The main spans, somewhat trimmed down temporarily from their 80-foot height, were lowered onto barges and floated down the Ohio to the back channel of Neville Island, which they thenceforth connected with Coraopolis. In 1994, they were scrapped at last because their steelwork was regarded as too brittle for safety.

Theodore Cooper, a distinguished engineer and a very innovative one, came eventually to a bad end. In 1907 he was consulting engineer on the Quebec railway bridge, which was to have an 1,800-foot main span of as light a construction as possible. His ignoring of telegrams, from the site warning of deflections in one cantilever, and administrative muddle once he realized the potential for trouble permitted the collapse of the cantilever with the death of 80 men. Blame circulated freely, but Cooper's career was over.

Off to Neville Island, with the top chord temporarily removed.

Over the back channel at Neville Island, 1927–94.

Fort Wayne Bridge

Allegheny River at Mile 1, between Pittsburgh and Allegheny (North Side)

1903; raised in 1918

Built by the Pennsylvania Railroad

The predecessor of this bridge, built for the Pennsylvania-leased Pittsburgh, Fort Wayne & Chicago Railway, had delicate wrought-iron latticework of a sort wholly supplanted in this new work by a main span of about 335 feet: four Pennsylvania trusses, elaborations of the Pratt truss, executed in heavy steel and accompanied by four side spans of Pratt trusses. The overall length was 965 feet. This was a construction to withstand what a modern locomotive had become: an advancing mass of steel, iron, water, and coal, weighing perhaps 250,000 pounds, pounding the rails with every revolution and pulling heavy loads of whatever freight was to be hauled. The bridge was double-decked, and though each deck was three tracks wide it held only two, set off-center.

In 1903 the river pool level at the Point and up beyond the Fort Wayne Bridge was 703 feet above sea level, and the bridge as built was no more of an obstruction to steam navigation than the Union Bridge of 1874 and the Sixth Street Bridge of 1892, downstream. But the Union Bridge came down in 1907, to be replaced by the Manchester Bridge with a much greater vertical clearance, and there seems to have been agitation to replace the Sixth Street Bridge in the same way. The Fort Wayne Bridge was raised in 1918, apparently at the insistence of the Secretary of War, so that the elevation of "low steel" became 750.9 feet, about 48 feet above river level; the concrete added to the stone piers is conspicuous. The subsequent raising of the Emsworth Dam has lessened the vertical clearance, though, to about 41 feet.

The bridge as built (above). **The bridge after 1918,** with the Sixteenth Street Bridge in the background (right).

Fort Wayne Bridge, 1867 – c. 1903

Felician Slataper, engineer

Privately built

The oldest rail line into Pittsburgh, by a brief margin of time, was the Ohio and Pennsylvania Railroad, which by 1856 had joined with two other Midwestern railroads to become the Pittsburgh, Fort Wayne & Chicago Rail Road Company. The Pennsylvania Railroad—just a Pittsburgh-Philadelphia operation then—had a large investment in this westward line, and a wooden railroad bridge was built in 1857 across the Allegheny River to connect the two lines. This was replaced a decade later by the wrought-iron lattice-truss bridge shown here. This had five 178-foot spans. By this time, incidentally, the Pittsburgh, Fort Wayne & Chicago Rail Road Company had been reorganized as the Pittsburgh, Fort Wayne and Chicago Railway Company.

This is remarkably lighter in construction than the third Fort Wayne Bridge, its successor.

S. V. ALLBEE, Photographer.

GEO. W. CASS,
President.

J. B. JARVIS,
Consulting Engineer.

Total Length, 1172 feet.

H. A. GARDNER,
Chief Engineer.

FELICIAN SLATAPER,
Engineer in Charge.

Wrought Iron Bridge, "Pittsburgh, Fort Wayne and Chicago Railway," over the Allegheny River,

AT PITTSBURGH, PA.

Sixteenth Street Bridge

Allegheny River at Mile 1.3

1923
H. G. Balcom, engineer;
Warren & Wetmore, architects

Built and owned by Allegheny County

The present bridge stands on the site of a wooden covered bridge, built in 1837, burned and rebuilt in 1851, partly washed away in 1865, and burned in 1918. By the last year, the U. S. War Department had come to insist on vertical clearances allowing steamboat navigation on the Allegheny, and a wholly new type of bridge was required. Another body, too, had come to affect bridge design in Pittsburgh: the Art Commission of the City, which had the right to dictate the general appearance of the bridges within its limits. Probably their influence was behind the County's hiring of the New York architects Warren & Wetmore, best known for having

View from shore level.

Sixteenth St. Bridge

finished Grand Central Station, to provide the general design that the New York engineer H. G. Balcom turned into a structure. The division of the bridge into three spans of heavily framed tied arches, one of 437 feet between two of half that size, is surely an "architectural" solution; the same is true of course of the nonstructural stone piers at the entrances, with their delicate carving and the bronze horses and armillary spheres by the New York sculptor Leo Lentelli. There are bridges in Paris, St. Petersburg, and elsewhere that attempt such marriages of engineering and "architecture," the latter conceived as decorative construction, and though our one specimen looks a little false, it is also a very likeable bridge. From an urban design viewpoint its 1,900-foot length connects not very much to not very much and seems to be patiently awaiting plazas, monuments, and fountains to terminate its axis.

August 19, 1922: the central arch under construction (above).

September 28, 1922: one of the shore spans nearly completed, but with the falsework still in place (right).

A trial maquette for the armillary spheres.

An armillary sphere, with horses, by Leo Lentelli.

Düsseldorf, 1898: another "architectural" bridge with attempted Renaissance character.

Washington Crossing Bridge

Allegheny River at Mile 3.2

1923
Benno Janssen, architect; Charles Davis et. al., *engineers*

Built by Allegheny County
Present owner: PennDOT

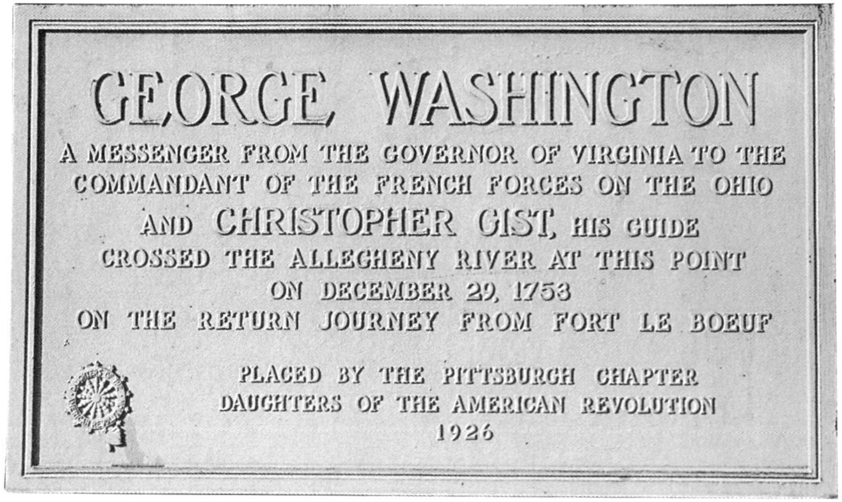

In this instance we name the architect first, because it was to him that the County Commissioners, under the influence of the Pittsburgh Art Commission, entrusted the design. In this regard it was similar to its contemporary, the Sixteenth Street Bridge, and such was the rage of the local engineers that it never happened again; thenceforth the architect had a contribution to make, but his work on County bridges was definitely subordinated to that of the engineers.

The dry information on the Washington Crossing (or Fortieth Street) Bridge gives its length over all as 2,366 feet, with three steel deck-arch spans, the main one being 360 feet. (The main spans of the Smithfield Street and Panther Hollow Bridges are also 360 feet; curious, how effects of scale can vary despite the coincidence of a uniform dimension.) The architects combined simple but carefully shaped piers and solid-web steel arches that were like curved plate girders, with flanges projecting

Looking across to Millvale in 1934. The three solid-web arches of the main spans.

The Washington Crossing Bridge: an architect-designed bridge.

radially from their sides. To these simple elements were added some lettering on the abutments, sculptures in the keystone positions, and escutcheons with armorial bearings on the railings. Opinion in the 1920s seems to have been that this bridge was an artistic success, while the Sixteenth Street Bridge was considered a piece of foolishness.

Just downstream is the Thirty-first Street Bridge of 1928, similar in overall form but unadorned and never greatly noticed. Again, it has a 360-foot main span, but a total length of 3,000 feet.

Tributes to an historic event (above). The bridge commemorates George Washington's near-fatal winter crossing of the Allegheny River by raft.

The County escutcheon, concealing a hinge (right).

The seals of the original thirteen states, polychromed years ago, are showing signs of deterioration.

Ohio River Bridges

West End Bridge

Ohio River above Mile 1

1930–32
Allegheny County Department of Public Works, engineers;
George S. Richardson, principal designer

Built by Allegheny County
Present owner: PennDOT

To some Pittsburghers, this is the handsomest of all local bridges, especially in view of its siting. The structure is almost devoid of decoration, but the lightness and good proportions of this bold arch at the head of the Ohio River give the old epithet "Gateway to the West" an air of literal truth.

The bridge is 1,891 feet in overall length. From north to south, there are four Warren pony trusses, the main span that has a 778-foot tied arch, and—originally—three Warren pony trusses. (To explain: a Warren truss is composed wholly of equilateral triangles, and in a pony-truss bridge the upper chords are not connected.) The main span has a stiffening truss concentric with the arch itself. The floor structure is suspended from wire cable hangers, which were subjected to the process of pretensioning, novel in the United States around 1930, so that the floor structure would not sag under anticipated loads. The tied-arch form was dictated by navigational clearance requirements, and the bridge probably was the largest to use this form at the time. When it was built, the vertical clearance was 73 feet, and today is 66.

Officially named the West End-North Side Bridge, here it is shown when new. To the left, smoke rises from the now-gone Reed Power Station on Brunot's Island.

Under construction in 1931. At the far end on the North Side, the stacks of the LaBelle Works of Crucible Steel are visible.

Three Warren pony truss approach spans.

McKees Rocks Bridge

Ohio River at Mile 3.3

1930–32
Allegheny County Department of Public Works, engineers;
George S. Richardson, principal designer

Built by Allegheny County
Present owner: PennDOT

Carl Condit, historian of "American building art," finds the McKees Rocks Bridge of "most unsatisfactory appearance": not because the individual spans are unpleasant to see but because there are so many kinds. From west to east, there are two through crescent arches of 300-foot span each; 1,145 feet of terra firma; eight deck trusses, about 163 feet each; one cantilever deck truss of about 194 feet; one cantilever deck truss of 340 feet; a river span that is a two-hinged trussed arch of 750 feet; three cantilever trusses of 338, 324, and 334 feet; and a humble 77-foot deck girder, for a total length with approaches of almost 7,300 feet.

The delicate strength of the main span.

The McKees Rocks Bridge when new; in the foreground, a Victorian Pratt truss.

The river span though, which resembles in general outline the 978-foot Hell Gate Bridge arch in New York, is much lighter and more delicate as befits a bridge that carries road vehicles rather than trains. Both arches have upper chords leveled out at their ends to make a graceful meeting with decorative towers.

In the early days, a traffic circle connected Ohio River Boulevard with the McKees Rocks Bridge.

Two crescent arches over the railroad at McKees Rocks (left).

An approach span at the east end of the bridge: September 10, 1930 (bottom left).

Sculptured ornamentation by Frank Vittor on one end pier of the main span (bottom right).

Sewickley Bridge, 1911 – c.1980

Ohio River near Mile 12

James G. Chalfant, engineer

Built by Allegheny County

Some time in late 1910 or early 1911, in the first view shown here, the Sewickley part of the Sewickley-Coraopolis Bridge is reaching out toward its similarly extended counterpart to the south. The main span will be 750 feet over the Ohio River, the whole bridge 1,853 feet long. Of this bridge, by the County Engineer James G. Chalfant, Carl Condit has said, "the bridge . . . was inordinately heavy. It was the greatest dead weight of any cantilever in the United States except for New York's Queensboro Bridge, an awesome, if dubious, distinction." It came down around 1980, replaced by a continuous-truss bridge of the same proportions. Two of the tower finials remain, one at Sewickley, one at Station Square in Pittsburgh.

The Sewickley truss, reaching out.

Equipped for trolley traffic, in a view of 1936.

The Sewickley Bridge of 1911 in its later days.

The demolition in 1980: the suspended span is being lowered.

A finial, now preserved, at Station Square.

Pittsburgh & Lake Erie Railroad Bridge

Across the Ohio River at Beaver, at Mile 25.6

1908–10
Albert Lucius, consulting engineer

Built by the Pittsburgh & Lake Erie Railroad

We have entered our neighbor Beaver County to examine a formidable cantilever bridge, built for the Pittsburgh & Lake Erie Railroad. This connected the main line through Pittsburgh, with the Pittsburgh & Lake Erie Railroad's principal station, and McKees Rocks, with its shops, to the principal trackage extending toward the Great Lakes.

The bridge is actually built as a pair. The 1,409-foot cantilever element has a channel span of two cantilevers and a suspended span, totaling 769 feet. At the end of one 320-foot anchor cantilever span is a more or less conventional truss bridge of 370-foot span. The main spans are 1,779 feet long and a little over 35 feet wide. The total cost was $1,550,000.

The bridge was both admired and criticized around 1910, the latter because the use of eyebars and pin connections—readily seen at sidewalk level on the Smithfield Street Bridge—was regarded as rather old-fashioned.

A suspended span under construction: October 19, 1909 (opposite).

The bridge in 1998 (above).

A file of locomotives testing the bridge: May 10, 1910 (below).

STEEL FURNISHED & ERECTED BY
McCLINTIC-MARSHALL CONS. CO.
NEW YORK PITTSBURGH CHICAGO.

Inland Bridges

Diamond Market, c. 1914 – 61

Diamond Street (Forbes Avenue) and Market Street

Rutan & Russell, architects

Built by the City of Pittsburgh

Following a northern Irish practice, what we now call Market Square was originally called the Diamond. Since 1783 it has received a variety of configurations, and the burning of an earlier structure in 1912 led to the most novel of the lot: a three-story double structure, its elements connected by a third-story footbridge over Market Street, and with each half pierced by a two-story archway allowing Diamond Street to pass through. The result, seen from our viewpoint, is a pair of bridge-like structures connected by yet another bridge. The spans of the openings were 50 feet, and each element was 168 by 68 feet, approximately, and 50 feet high.

The Market around 1920: red brick, white terra cotta. Market Street is in view; Diamond Street passes through the arches to the left and right (above).

The Market in its last days, around 1960. In view, an arch through which Diamond Street passes (right).

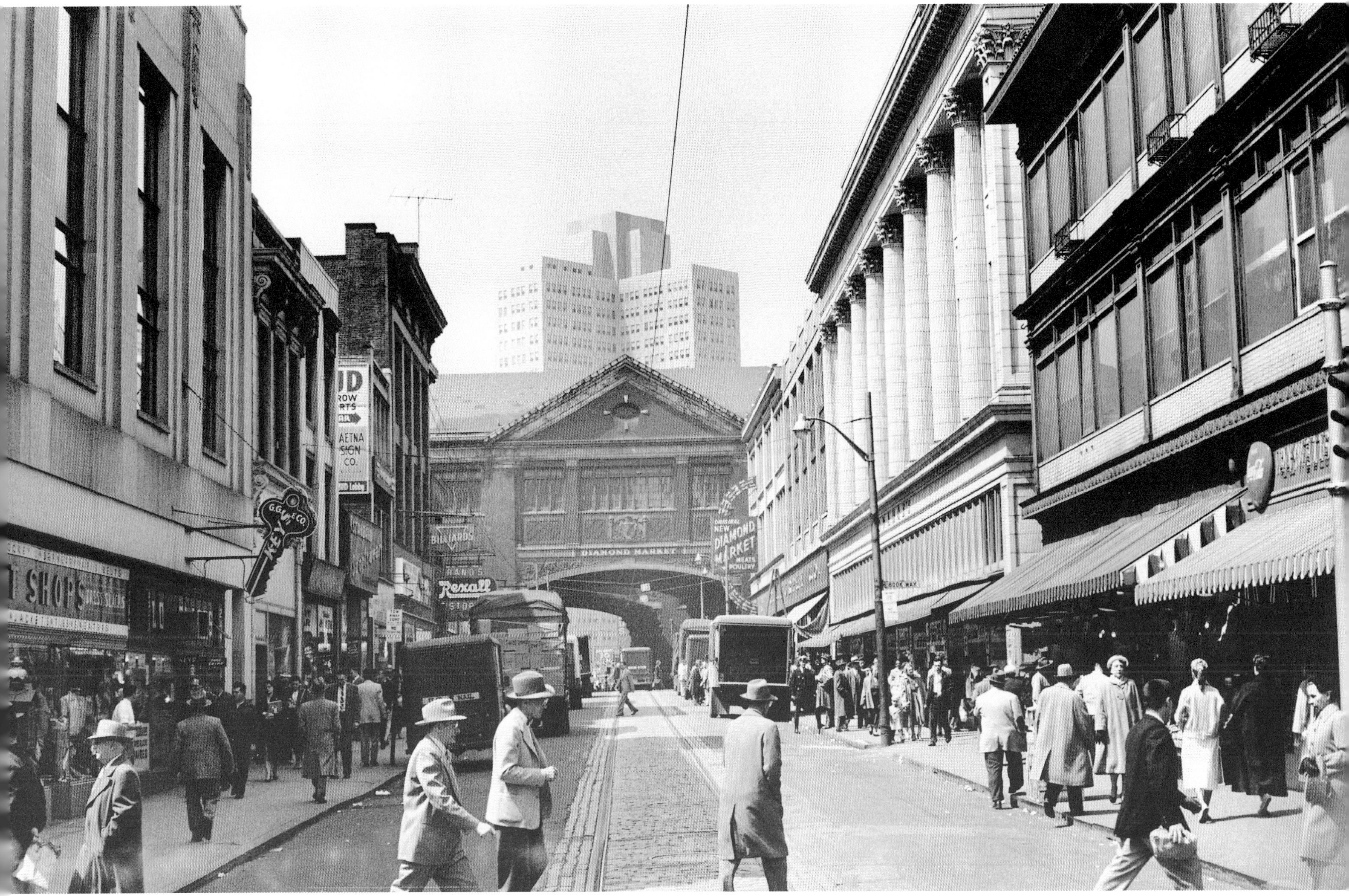

Diamond Street, later Forbes Avenue.

"Bridge of Sighs," Allegheny County Buildings

400 block, Ross Street, Pittsburgh, between the Allegheny County Courthouse and Jail

C. 1888
Henry Hobson Richardson, architect

Built and owned by Allegheny County

This pedestrian bridge is fairer to see from without than from within, for it was built for the walks of prisoners between courts and the Jail. It retains the Romanesque character of the buildings at either end, and the Jail end is furthermore surmounted by a reduced-scale version of the tower on some medieval bridge that might have stood ready to obstruct traffic with a portcullis. Yet the Pittsburgh bridge has also a sort of pedimented aedicule at its midpoint; this is a Renaissance rather than a medieval feature, recalling the Venetian Rialto Bridge.

The well-proportioned, rock-faced segmental arch has a span of 52' 6". The opening through the archway is about six feet higher than it once was, due to a lowering of Ross Street early in this century.

The facing of the bridge, like that of the Courthouse to the left, is tied into the backup brickwork, with low courses that enter the brickwork and support the higher and shallower courses (right).

H. H. Richardson's Allegheny County Courthouse and Jail, connected by the Bridge of Sighs (opposite).

The Pont Valentré, Cahors, France. A similar composition from the Middle Ages, with a defensive tower at the end.

The Jail in recent times, before its adaptation in 1999 for family and juvenile court facilities.

The bridge in the 1890s, before the lowering of the "Hump" brought down the level of Ross Street (opposite).

TRADES JOURNAL

Mercy Hospital Footbridge

Locust Street, the Bluff

1905
Edward Stotz, architect

Built and owned by Mercy Hospital

As built, this bridge connected the main hospital building with the New Nurses' Building. In 1988 the former was demolished but the terra-cotta siding of the bridge was preserved, and around 1995 the bridge was reconstructed, moved, and lifted into place around 200 feet further along Locust Street. It now connects the New Ambulatory Care Building with a parking garage.

The structure is steel trusswork, bearing a terra-cotta cladding with Gothic detailing and a Spanish clay-tile roof. The span is about 55 feet. As rebuilt, the bridge has lost some terra cotta, so as to admit more natural light and expose some of the inner steelwork.

The footbridge as rebuilt and moved. The detailing recalls the 15th century in northern Europe, a time and a place where, in fact, terra cotta was not a major architectural material.

Mount Washington Roadway Bridge, 1928 – 83

Over the "saddle," Mount Washington

Charles M. Reppert, chief engineer; Stanley L. Roush and J. Vernon Wilson, architects

Built by the City of Pittsburgh

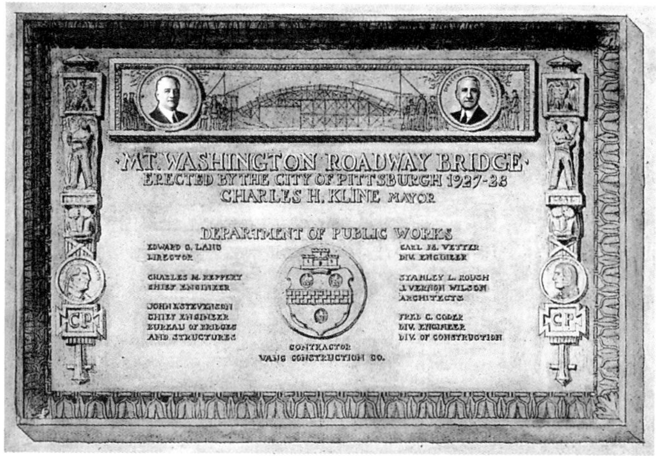

In 1983 we lost this concrete bridge of the Mount Washington (or P. J. McArdle) Roadway, styled by Stanley L. Roush and built in 1928. A bronze tablet was designed for the bridge that shows, among other things, how the falsework supported the arch, whose span was 203 feet. Roush's services appear to have been on loan from the County for this City project, in a collaboration with J. Vernon Wilson, a Bellevue architect.

Plaque designed by Stanley Roush (above).

The deck being laid, and period roadmaking machinery: June 19, 1928 (right).

The bridge when new, in 1928.

The Mount Washington Roadway Bridge complete; beyond, the clouds and vapors of the South Side.

East Street Bridge

Over I-279, about 1.5 miles north of East Ohio Street, Pittsburgh, connecting North Charles and Essen Streets

1930

Built and owned by the City of Pittsburgh

Here is a graceful feature of a once-picturesque valley, a majestic hello or goodbye from the Pittsburgh area in much the same way as the Westinghouse Bridge is. It is an arched cantilever, with a deck about 225 feet above the East Street that once gave the valley and the bridge their names. The total length is about 1,200 feet with a main span of about 500 feet.

The East Street Bridge is sometimes known as the Essen Street or E. H. Swindell Bridge.

A contrast in scale.

An early view of one of the hinges (left).

The East Street Bridge: from the valley floor (below).

The center span of the East Street Bridge nearly closed.

Sewickley Creek Bridge

Over Sewickley Creek, linking Little Sewickley Creek and Woodland Roads, Edgeworth

1889
Charles Davis, engineer

Built and owned by Allegheny County

Charles Davis was County Engineer in the 1880s. He was once quoted as implying that H. H. Richardson's Allegheny County Courthouse and Jail were unnecessarily expensive. Here, he has created a nice little work in stone that fits well into the rising, wooded landscape. The upward curve of the roadway speaks of a horse's slow advance, not an impetuous motorized dash to the country house.

The quiet craftsmanship of the parapet (right).

The bridge bows out as the road arcs upward (opposite).

Atherton Avenue Bridges:

Baum Boulevard over the Pittsburgh Junction Railroad, 1913 – 1973

Baum Boulevard over the Pennsylvania Railroad, 1913 – c.1985

Built by the City of Pittsburgh

Bureau of Construction progress shots. The architect Stanley Roush was probably involved; his contemporary portal for the Corliss Tunnel shows an arch like the main one here: almost a segmental arch, but rounded off at the springings.

Here is the progress of the Atherton Avenue (i.e., Baum Boulevard) Bridge over the Pittsburgh Junction Railroad, recorded on June the 26th and September the 19th, 1912, and January the 2nd, 1913. The main span is 168' 6". The Pittsburgh Junction Railroad left the Baltimore & Ohio in Hazelwood near the Monongahela, passed northwest up Junction Hollow along the west side of Schenley Park, entered a tunnel under Neville Street just north of Forbes, emerged into the light just north of Centre, crossed under the Pennsylvania main line and ran parallel for a while, then crossed the Allegheny River and Herr's Island at Thirty-third Street to join another Baltimore & Ohio line on the right bank.

B. OF C. ATHERTON AVE. BRIDGE. OVER P.J.R.R. GENERAL VIEW LOOKING NORTH.
1-2-13 3087

About 200 feet to the east was another Atherton Avenue Bridge, this one crossing the Pennsylvania Railroad main line. Instead of the unequal triple arch spans of the bridge over the Pittsburgh Junction Railroad, there were three 90-foot arches that dispensed with the formidable wooden centering in favor of light steelwork. In the earlier view from September 19, 1912, a great spray of reinforcing rods rises from a completed pier. This, apparently, will be bent down over the adjoining centerings just before the concrete is poured. The second view shows the bridge on February 14, 1913, not long before completion.

The bridge over the Pennsylvania main line was less ambitious but went around 1985 (above and right).

The final moment for the Pittsburgh Junction Railroad bridge, 1973 (opposite).

Bellefield Bridge

Schenley Plaza, Schenley Park

1897
Henry B. Rust, engineer

Built by the City of Pittsburgh

***A Song to Nature* by Victor Brenner,** 1918, on top of the buried bridge.

This bridge and St. Pierre Ravine, the void it spanned, were engulfed in earth around 1913: spoil, it is said, from the Grant's Hill Hump lowering in the eastern part of the Triangle. Schenley Plaza now lies along the buried little ravine, and the Bellefield Bridge is the foundation for the Schenley Fountain.

The bridge was to serve as a link between Oakland and the Schenley Bridge, also a Rust design, that was to replace a fire-damaged bridge in the same place. The Bellefield Bridge had a single arch of gray sandstone, backed with concrete, with a span of 150 feet. The total bridge length was 341 feet, with an 82-foot width. St. Pierre Ravine at this point was about 70 feet deep. A report of the time said that William Falconer, a landscape architect and the park superintendent, was going to transform the gash-like ravine into a lake.

The bridge as completed. Hotel Schenley is in the background.

The Bellefield Bridge and the Carnegie Institute in its original form.

Panther Hollow and Schenley Bridges

Over Junction and Panther Hollows, respectively, Schenley Park

1897
Henry B. Rust, engineer

Built and owned by the City of Pittsburgh

One of Giuseppe Moretti's panthers.

The Panther Hollow Bridge.

The burning of the Casino at the entrance to Schenley Park in 1896 and serious damage to the new Schenley Bridge near by seem to have been the occasion for the building of *two* steel arch bridges of the same dimensions: the new Schenley Bridge crossing Junction Hollow between the Carnegie Institute and Phipps Conservatory, the other across Panther Hollow.

The Panther Hollow Bridge is the better known in the engineering literature: abutments of rock-faced sandstone ashlar with 28-foot openings that have elliptical arches, and three-hinged steel parabolic arches, four abreast, with a 360-foot span and a 45-foot rise, the ravine being about 120 feet deep. The Panther Hollow Bridge is

The Schenley Bridge, with decorative railings of the 1930s and chain-link fencing installed in 1995.

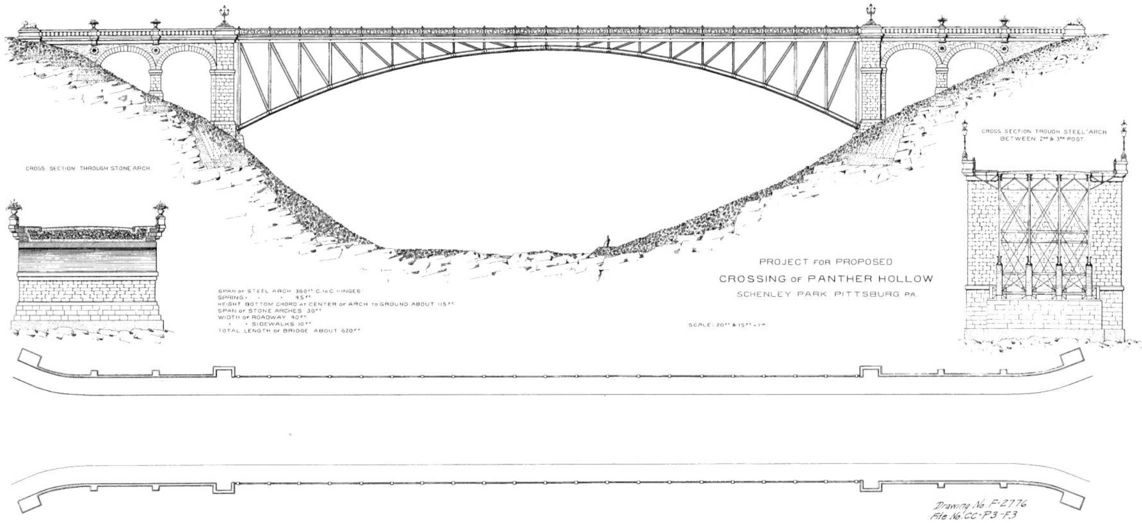

A preliminary design for the bridges.

distinguished by four bronze panthers by the local sculptor Giuseppe Moretti. The Schenley Bridge had no such distinction, but in connecting the Carnegie Institute to the Hornbostel buildings at "Carnegie Tech" and the Phipps Conservatory not far away, it was an element of a gracious ensemble of architecture and landscape. But in the last few years the Muse of Liability has blurred the views with chain-link fencing, and the majestic Hornbostel building that rose from the Hollow's wild slope has been half-hidden by a new building, of the 1990s but 1960-ish in looks.

The first Schenley Bridge, with a trolley trestle in the nearer distance; a view of 1893 (opposite).

The Schenley Bridge from Schenley Drive (below).

Tufa Bridges

Bridle Path between Bartlett Street and Serpentine Drive, Schenley Park

1908
Pittsburgh Department of Public Works, engineers

Built and owned by the City of Pittsburgh

George W. Burke, Pittsburgh's Superintendent of Parks in 1908, had to carry the Schenley Park Bridle Path over two voids at the upper end of Panther Hollow. His solution was a pair of reinforced-concrete bridges, conventional except that tufa from Ohio was stacked inside the formwork and thus became a facing. Tufa is a very white, lumpy, porous material, cryptocrystalline calcium carbonate deposited in streams. Visitors passing through the Phipps Conservatory also have a miniature tufa bridge along their path.

One of the bridges today, with the Panther Hollow Bridge in the background (above).

Two views in the early days (right and opposite).

Beechwood Boulevard Bridge

Beechwood Boulevard, connecting Greenfield Avenue with Schenley Park

1921–23
Charles M. Reppert, chief engineer; Stanley L. Roush, architect

Built and owned by the City of Pittsburgh

This bridge, also called the Greenfield Bridge, deserves celebration for what it *was* more than for what it is. An alteration in 1980 reduced decoration on this once-ornate bridge to its entrance pylons and its main arches, removing the grand series of sculptured lamp standards and urns by Stanley Roush. A simpler composition might have been as effective; the effect of so many features was a little disunified perhaps. Yet this *was* an approach to Schenley Park, and a little fanciful swank was not inappropriate. The bridge now is a degraded leftover, surely.

The main span is 279 feet, and the total length is 426 feet.

The bridge under construction (above), and as completed (opposite). There was no main-traveled road passing beneath; the urns and obelisk-like lamps lined a stately back entrance to Schenley Park.

BEECHWOOD
12/23/23 7068

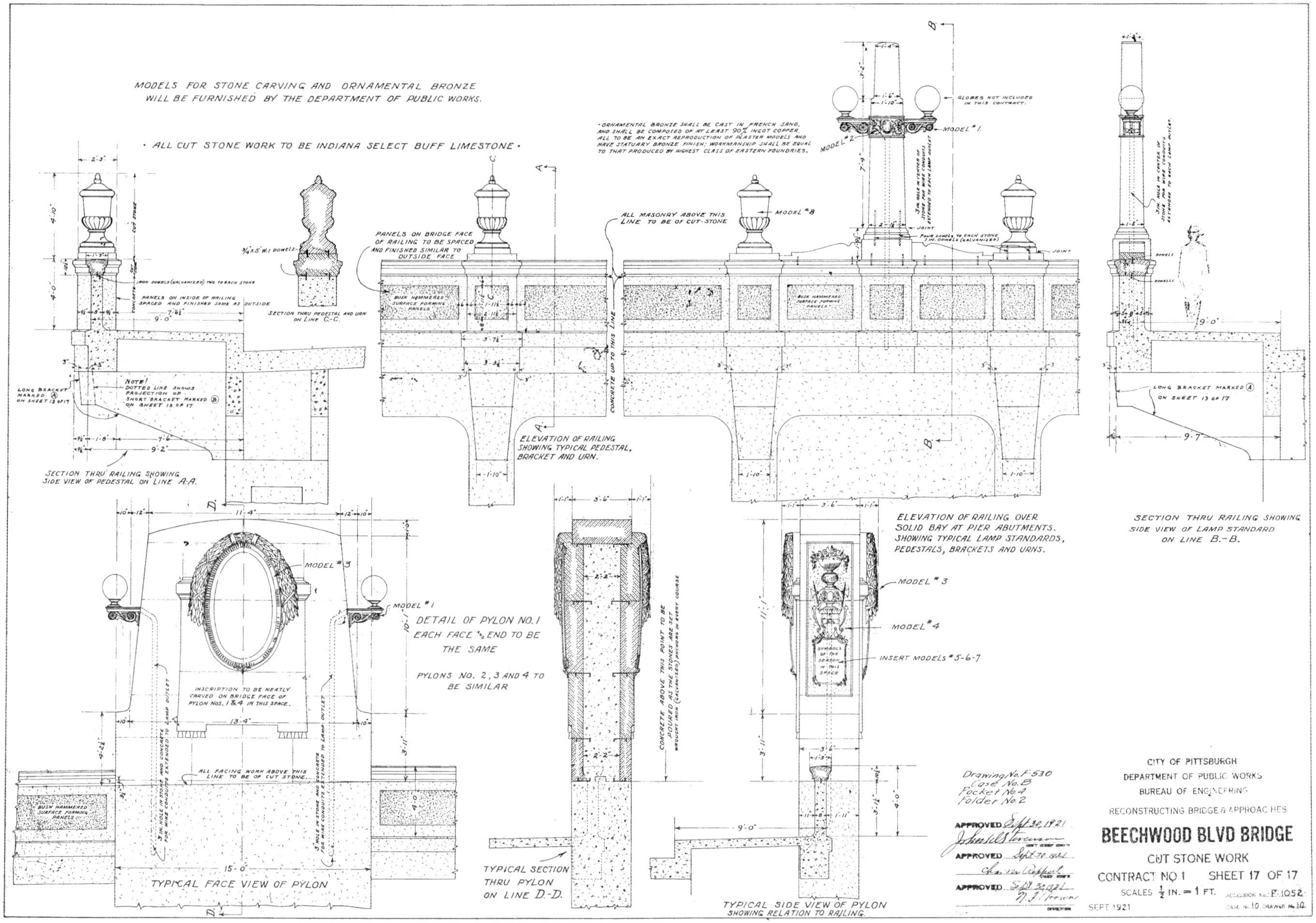

A drawing by Stanley Roush.

A stela left untouched during the stripping of 1980.

The Beechwood Boulevard Bridge at deck level as it looks now (above).

The 1980 reconstruction and stripping (above) with most of the decorative work removed and new lampposts (right).

Heth's Run Bridge

Butler Street at the western border of Highland Park, Pittsburgh

1914
T. J. Wilkerson, engineer; Stanley L. Roush, architect

Built by the City of Pittsburgh
Present owner: PennDOT

The Heth's (or Haight's) Run Bridge dates from 1913 and was an early beneficiary of Stanley Roush's architectural services. The heavy balustrade with dwarf piers and urns looks a little uneasy cantilevered on cyma-reversa brackets, but the detailing itself is quite handsome. The City, having gone to such trouble with this bridge on the western edge of Highland Park, decided to use the ravine as a trash dump, and even today the bridge is nearly buried.

Details by Roush (right), and the bridge when new (opposite).

B. OF E. GENERAL VIEW NO. 1 HAIGHTS RUN BRIDGE.
6-1-15 4465

What Heth's Run Bridge has come to.

Larimer Avenue Bridge

Larimer Avenue over Washington Boulevard, Pittsburgh

1912
T. J. Wilkerson, engineer, Pittsburgh Department of Public Works

Built and owned by the City of Pittsburgh

The bridge today, after floor reconstruction.

Among concrete arches, this bridge had the second greatest span in the world when built: 300' 4". It is 670 feet long and 50 feet wide, and crosses a valley about 100 feet deep. A purist may object to the presence of masonry imitations here: rustications on the piers and abutments, arches with capitals on the spandrels. One might have wished a more natural expression for reinforced concrete. Yet a look at the construction shots is most impressive; a heroic work is in progress, falsework, formwork, and reinforcing bars going all ways in a manner that must have defied coordination.

Missing now are the irrational but somehow very appealing lamps that the architect Stanley L. Roush designed for the piers: rostral columns with lamps perched on the rams of ancient warships, while eagles with raised wings rise on other lamps teetering on the column tops: the whole composition looking much better than it sounds.

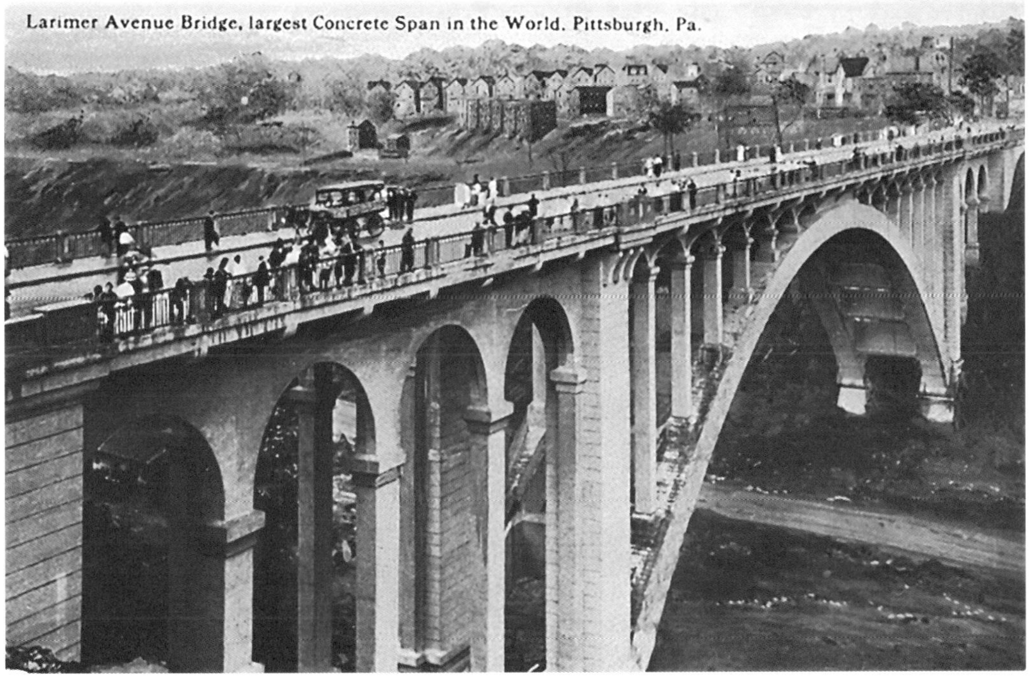

The falsework in place: August 21, 1911.

Main arches and spandrel columns nearing completion: February 10, 1912.

B. OF C. LARIMER AVE. BRIDGE. GENERAL VIEW FROM PAULSON AVE. 6-26-12 270

Now gone. Sometimes art can be good even when a little ridiculous.

The bridge and the billboard: two giants of their sorts (opposite).

The Brilliant Cutoff Viaduct (right) and the Lincoln Avenue Viaduct to its left.

Brilliant Cutoff Viaduct and Lincoln Avenue Viaduct

Washington Boulevard north of Frankstown Avenue

Brilliant Cutoff Viaduct, 1902
William H. Brown, engineer
Built by the Pennsylvania Railroad

Lincoln Avenue Viaduct, 1905
City of Pittsburgh Bureau of Construction, engineers
Built and owned by the City of Pittsburgh

Several circumstances may have impelled the Pennsylvania Railroad toward a seemingly retrograde policy of building heavy masonry bridges late in the 19th century. The prospect of heavier and heavier steam locomotives, compact masses of steel, cast iron, water, and coal, moving rapidly and pounding against the rails, must have made bridges of heavy, inert masonry, equal to anything the future might bring, seem investments well worth the cost. The Johnstown Flood, too, would have demonstrated the stresses that masses of moving water and heavy debris could put on anything in their way. Thus, when the Brilliant Cutoff of the Pennsylvania was constructed between East Liberty Station and Aspinwall, only the bridge across the Allegheny River,

The Brilliant Cutoff Viaduct.

The Brilliant Cutoff Viaduct.

The Lincoln Avenue Viaduct as it is today.

The Lemington Avenue arches of the Brilliant Cutoff Viaduct.

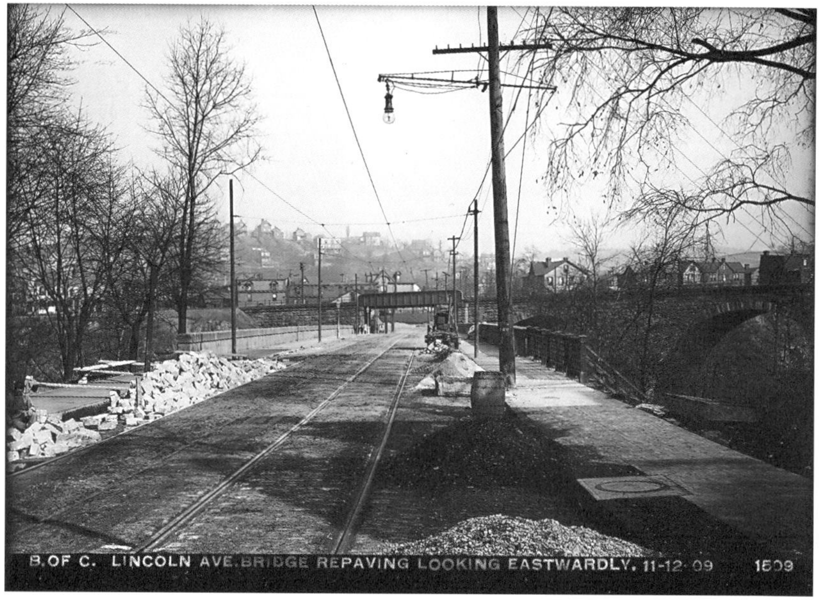

The Brilliant Cutoff Viaduct crosses Lincoln Avenue on a plate-girder bridge: a view from 1909.

with a channel span of about 400 feet, was composed of steel trusses. Elsewhere, the Cutoff passed over Lemington Avenue on one broad arch flanked by two narrower ones, and ran alongside Washington Boulevard on six arches of rock-faced ashlar, concrete, and rubble filling, five of 80-foot span, one of a hundred.

Expediency took over to the immediate north, and the Brilliant Cutoff crossed the right-of-way of the slightly newer Lincoln Avenue Viaduct on a plate-girder span. Lincoln Avenue ran northwest from Frankstown Avenue, and the decision to build its crossing over the Boulevard in two three-centered stone-faced arches of 130-foot span may have been the result of moral pressure from the handsome railroad viaduct and the park-like Boulevard itself, with the once-beautiful Silver Lake, favored for skating, just below.

What has happened to a once beautiful setting? The Lincoln Avenue Viaduct is to the left and the Brilliant Cutoff Viaduct is to the right.

Old Stone Bridge

Near Old William Penn Highway and Beatty Road, Monroeville

C. 1870

Privately built
Present owner: Municipality of Monroeville

Here is one more reminder that not all our bridges are dramatic crossings of ravines and rivers. A realignment of Route 22 in 1942 left this bridge to idleness. It was built to connect the former Patton Township with the old Philadelphia Turnpike.

The parapets have an almost organic character, rather earthworm-like.

George Westinghouse Memorial Bridge

Route 30 (Lincoln Highway) over Turtle Creek Valley

1930–32
Allegheny County Department of Public Works, engineers; George S. Richardson, principal designer

Built by Allegheny County
Present owner: PennDOT

Construction shots indicate the fact, at first startling, that the mighty-looking piers of this bridge were still incomplete when the main arches, and even some of the spandrel columns, were complete.

To a rail traveler coming from the east, the bridge is a welcome to Greater Pittsburgh, a triumphal arch and then some. The deck is 240 feet above the valley floor, and the main span beneath which the train passes is 460 feet across. Carl Condit is inclined to sniff at this, the longest span for a reinforced concrete arch in the United States—the Swedes have one of 866-foot span—but most of us, nonetheless, will be impressed. There is a peculiar beauty in the parabolic forms of the arches, which vary greatly in span and include two arches that are not complete but seem to plunge into the valley walls. The effect is one of uncompromised strength and

Frank Vittor's relief of *Electricity.*

The George Westinghouse Memorial Bridge and the Turtle Creek Valley. Note the ways in which the arches of the end spans seem to plunge into the valley walls.

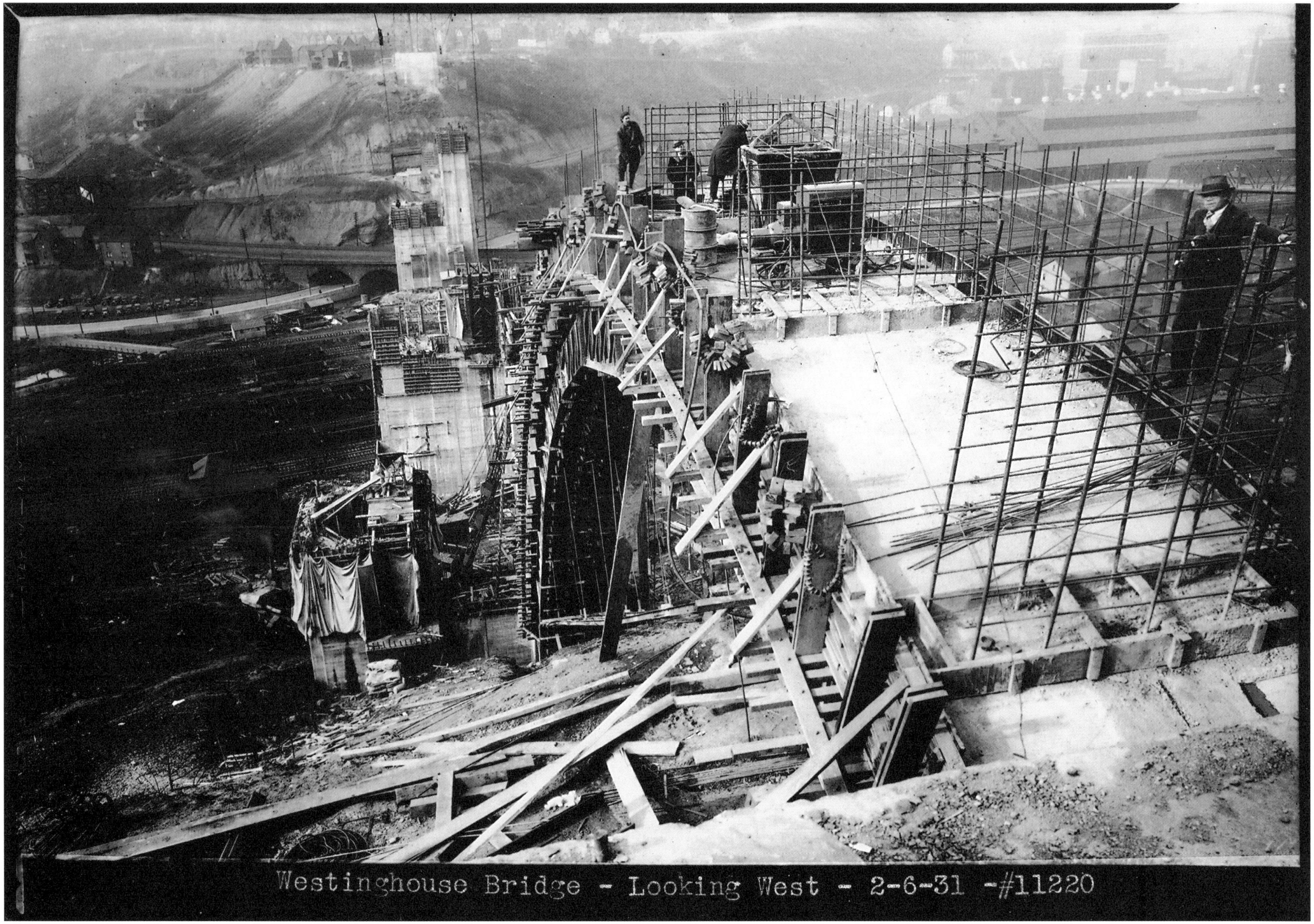

Early in the project, with formwork and reinforcement exposed.

simplicity. At road level the grandeur of construction is invisible, but four granite reliefs by Frank Vittor summarize, in a Modernistic way, the history of the Turtle Creek Valley.

Eventual deterioration led to a reconstruction campaign in 1982–83, with replacement and widening of the floor structure.

Opinions as to the length of the bridge vary, but 1,596 feet appears to be the likely figure.

Parabolic arches in place; spandrel columns being poured.

The bridge well under way, but with spandrels remaining to be built.

Frank Vittor's relief of *Steel* (above).

The George Westinghouse Memorial Bridge as eastern gateway: the motorist's and railroad passenger's views (above left and bottom).

Bridges and the Monongahela Wharf, May 9, 1932.

Further Reading and Investigation

Charette, The. 1920-71.

This was the magazine of the Pittsburgh Architectural Club, and thus a forum for professional opinion locally.

City Photographer Collection, Archives of Industrial Society, 400 North Lexington Avenue, University of Pittsburgh, Pittsburgh, Pa., 15260. (412) 244-7075. By appointment.

The collection includes excellent progress shots of public works of the City of Pittsburgh, from 1901 to 1958.

Condit, Carl W.

–*American Building Art: The Nineteenth Century.* New York: Oxford University Press, 1960.

–*American Building Art: The Twentieth Century.* New York: Oxford University Press, 1961.

These volumes are full of objective information.

Evert, Marilyn, and Vernon Gay. *Discovering Pittsburgh's Sculpture.* Pittsburgh, University of Pittsburgh Press, 1983.

This authoritative work includes information on the sculptures on the bridges, as well as brief biographies of the artists: Leo Lentelli, Giuseppe Moretti, and Frank Vittor, to name a few.

Pittsburgh District, U.S. Army Corps of Engineers

–*Allegheny River Navigation Charts,* various editions. $7.00

–*Monongahela River Navigation Charts,* various editions. $8.00

–*Ohio River Navigation Charts,* various editions. $8.00

These are good bridge-recognition books. They can be purchased at Room 2032, Federal Office Building, 1000 Liberty Avenue, Pittsburgh, Pa., 15222. (412) 395-7502.

Pittsburgh Photographic Library, Pennsylvania Room, Carnegie Library of Pittsburgh, 4400 Forbes Avenue, Pittsburgh, Pa., 15213. (412) 622-3154.

This is the most accessible collection locally.

Plowden, David. *Bridges: The Spans of North America.* New York: The Viking Press, 1974.

Plowden is basically a photographer, but his book is well researched, and Pittsburgh bridges have long been a favored subject.

Richardson, George S. "History of Allegheny County Bridges." MS., 1979.

Copies of this interesting memoir by a prominent local engineer are on file in the Pittsburgh History & Landmarks Foundation library. (412) 471-5808. By appointment.

Tyrrell, Henry Grattan. *History of Bridge Engineering.* Chicago: Henry Grattan Tyrrell, 1911.

This is an older source, preceding the big construction campaigns of Pittsburgh and Allegheny County.

White, Joseph, and M. W. von Bernewitz. *The Bridges of Pittsburgh.* Crafton, Pa.: Cramer Printing & Publishing Company, 1928.

Full of information, this was the last book devoted exclusively to Pittsburgh's bridges until the publication of this book.

The main span of the Seventh Street Bridge being constructed as cantilevers, with temporary struts. The shore spans are temporarily anchored. The year is 1927.

Index

Page numbers in boldface indicate major treatments of the subjects.

Setting the last keystone: the Bellefield Bridge, Schenley Park, 1897.

About the Author

Walter C. Kidney is the architectural historian of the Pittsburgh History & Landmarks Foundation, having joined the staff in 1985 after some years of freelancing. He has written a number of books for Landmarks, including: *Pittsburgh's Landmark Architecture* (1997); *Allegheny Cemetery: A Romantic Landscape in Pittsburgh* (1990); and *A Past Still Alive* (1989). Numerous articles by Mr. Kidney appear in *PHLF News,* Landmarks' membership newsletter. Among his other writings are *The Architecture of Choice: Eclecticism in America, 1880–1930* (1974) and *A History of the Pittsburgh Builders Exchange, 1886–1986* (1986).